General Biology I

Lab Manual

Werner Williams

St. John's River State College

Kendall Hunt
publishing company

Cover images © Shutterstock.com

www.kendallhunt.com
Send all inquiries to:
4050 Westmark Drive
Dubuque, IA 52004-1840

Contents

Lab 1: Tools for Scientific Investigation 1

Lab 2: Use of the Scientific Method 17

Lab 3: Atoms and Molecules 33

Lab 4: Solutions and pH...41

Lab 5: Detecting Organic Molecules...57

Lab 6: The Use and Care of the Microscope...67

Lab 7: Movement of Materials across Cell Membranes...................79

Lab 8: Enzymes 89

Lab 9: The Use of the Spectrophotometer...97

Lab 10: Cell Respiration...109

Lab 11: Cell Division...121

Lab 12: Human Genetics 131

<u>Lab 1: Tools for Scientific Investigation</u>

This lab activity is designed to assist you in using scientifically valid means of collecting, analzying, interpreting and presenting data in a clear format. A useful tool for taking measurements called the metric system will be discussed as well as using certain lab devices.

Scientific Notation

The speed of light is 300,000,000 m/sec and the weight of a particle of dust is 0.0   000 000 753kg. Scientist use **scientific notation** as a shorter way of writing such large (or small) numbers. It is based on powers of the base number 10.

The number 678,000,000,000 is written as 6.78×10^{11}.

The first number, which in this example is 6.78 is called the <u>*coefficient*</u> and must be <u>greater than or equal to 1 and less than 10</u>. The second number called the <u>*base* must always be 10</u>. The base number is always written with an exponent, and in the number 6.78×10^{11}, 11 is the <u>*exponent*</u> of the power of 10.

HOW TO WRITE IN SCIENTIFIC NOTATION:

1. Put the decimal after the first digit and drop the zeros.

    For the number: $678,000,000,000$ write as <u>6.78</u> .

2. To find the exponent, count the number of places from the decimal to the end of the number.

    In 678,000,000,000 there are 11 places, so it is written as: 6.78×10^{11}

    Another example: 29,000,000,000,000,000,000 is written as 2.9×10^{19}

3. For small numbers, we use a similar approach.

    For the number: 0.00000000000045 the coefficient is <u>4.5</u> .

4. Numbers smaller than one will have a negative exponent. To find the exponent count the number of places from the decimal to the end of the number.

    In 0.00000000000045 there are 13 places, so it is written as: 4.5×10^{-13}

5. Another example: 0.0001 is written as 1.0×10^{-4}

PRACTICE:

Write the following numbers in decimals.

1. $8.9 \times 10^{-7} =$

2. $1.66 \times 10^{4} =$

3. $7.2 \times 10^{-5} =$

4. $4.64 \times 10^{8} =$

Write the following numbers in scientific notation.

1. $0.0000032 =$

2. $71,000,000,000,000,000 =$

3. $0.33 =$

4. $96,000 =$

Metric Measurements

The metric system is used by scientists worldwide and is the official system of measurement in most countries. It is based on standard units that can be easily converted by simply multiplying or dividing by ten.

Table 1 Standard Units of the Metric System

Measurement	Standard Unit	Example
Length	Meter (m)	Height of a typical door handle (1m = approximately 39 inches)
Mass	Gram (g)	Mass of a one dollar bill (1 g = 0.035 oz)
Volume	Liter (l)	Volume of large fast food soda (1 l = approximately 1qt)
Temperature	Celsius (°C)	Water freezes at 0°C and boils at 100°C

Source: Werner Williams

We may also use a super-unit prefix or a subunit prefix to describe measurements. Super-units show multiples of the base unit and make the base unit larger. Sub-units make the base unit smaller.

Table 2 Common Metric System Prefixes and Their Value

	Prefix	Symbol	Value		
Superunit	**Mega**	**M**	**Million**	**1000000.0**	**10^6**
	Kilo	**k**	**Thousand**	**1000.0**	**10^3**
	Hecto	**h**	**Hundred**	**100**	**10^2**
	Deca	**da**	**Ten**	**10**	**10**
Unit	**Meter**	**M**	**One**	**1**	**1**
	Gram	**g**			
	Liter	**l**			
Subunit	**Deci**	**d**	**Tenth**	**0.1**	**10^{-1}**
	Centi	**c**	**Hundredth**	**0.01**	**10^{-2}**
	Milli	**m**	**Thousandth**	**0.001**	**10^{-3}**
	Micro	**µ**	**Millionth**	**0.000001**	**10^{-6}**
	Nano	**n**	**Billionth**	**0.000000001**	**10^{-9}**

Source: Werner Williams

Metric Conversions

Conversions within the metric system can be made easily using a metric staircase (ladder). Figure 1. Each step of the staircase represents a 10-fold change in the value of the measure or a shift of the decimal point one place. Therefore, each step moved down the staircase represents a movement of the decimal point one place to the right.

Each step up the staircase represents a movement of the decimal point one place to the left. Four steps up or down the staircase represents a movement of the decimal point four places to the left or right.

To remember the prefixes in order, at least the middle seven (from kilo to milli), just remember this phrase: **K**ing **H**enry **D**oesn't [**U**sually] **D**rink **C**hocolate **M**ilk. The **U**sually is left in the middle to represent the base "unit" (gram, liter or meter).

Kilo- hecto- deka- [unit]- deci- centi- milli-

Since each step is ten times or one-tenth as much as the step on either side, we have:
1 kilometer = 10 hectometers = 100 dekameters = 1000 meters
= 10,000 decimeters = 100,000 centimeters = 1,000,000 millimeters

Alternatively, we have:
1 milliliter = 0.1 centiliters = 0.01 deciliters = 0.001 liters = 0.000 1 dekaliters
= 0.000 01 hectoliters = 0.000 001 kiloliters

As we move from one prefix to another, the decimal point is moved one place, filling in the space with a zero. To move to a smaller unit of measurement (moving down the staircase), we move the decimal point to the right the same number of places and vice versa.

LADDER METHOD

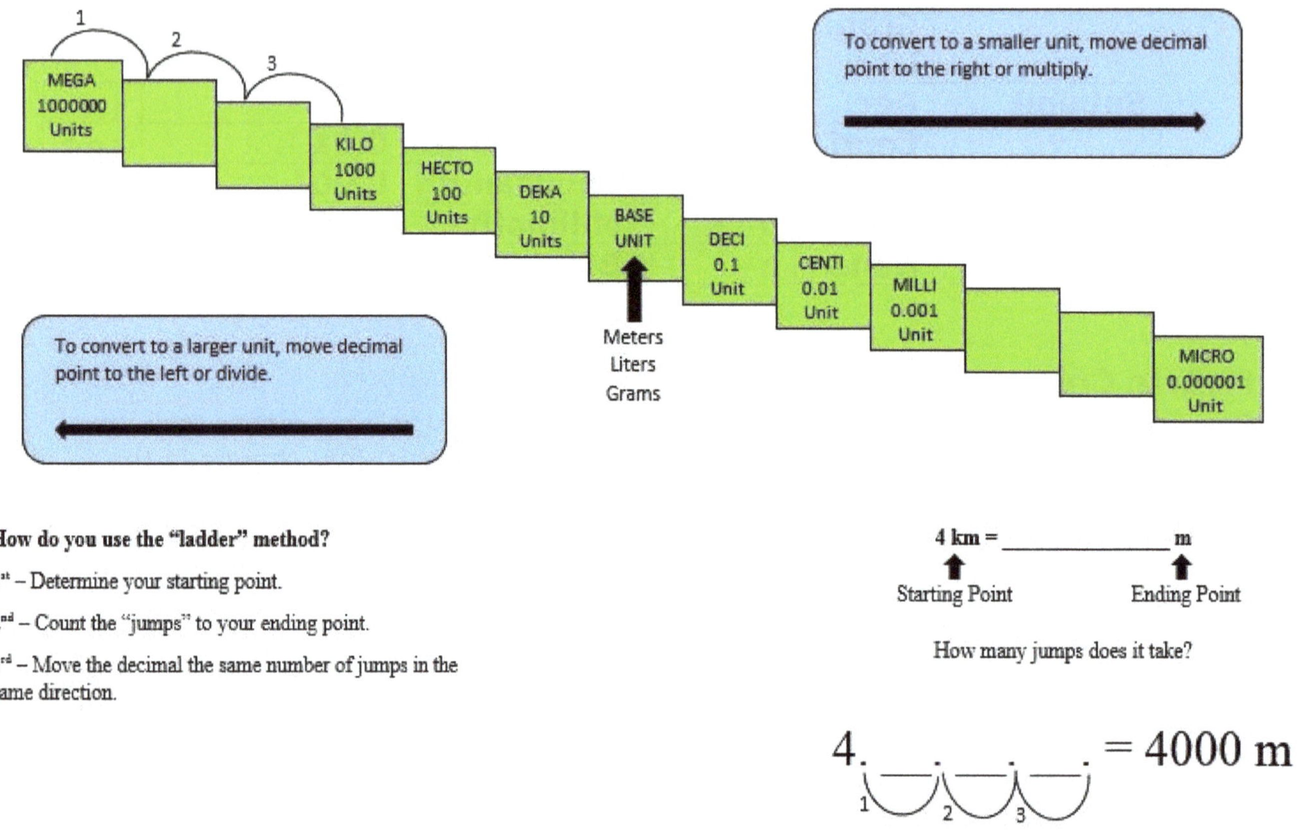

Contributed by Jennifer Raabe © Kendall Hunt Publishing Company

Figure 1 A Metric Staircase

Example: Convert 13.89 kilometers to centimeters.
How many jumps is it from "kilo-"to "centi" and in what direction?
Five to the right
Therefore, start with 13.89 and move the decimal point 5 spaces to the right filling in the space with zeros.   **1389000**

13.89 kilometers = 1389000 centimeters (or 1.389×10^6 cm)

Example: Convert 625 centigrams to hectograms.
How many jumps is it from "centi-"to "hector-" and in what direction?

Four to the left
Start with 625 and move the decimal point 4 spaces to the left filling in the space
with zeros.   **.0625**

625 centigrams= 0.0625 hectograms (or 6.25×10^{-2} hg)

PRACTICE:
Convert the following:
1. 1500m to km.
2. 600hm to mm
3. 60mg to g
4. 4kg to dg
5. 0.2 liter to ml
6. 2300cl to hl

Measuring Mass Using an Electronic Balance

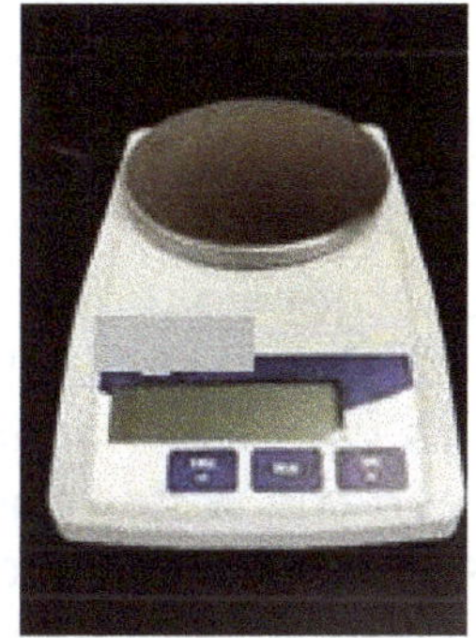

Source: Werner Williams

Figure 2 A typical Electronic Balance

A **gram (g)** is the basic unit for measuring mass in the metric system. About 454g
make a pound. An electronic balance is a common device found in biological labs
and is used for precise measurements. Chemicals should never be placed directly
on the balance pan, therefore use a container or paper. The paper is called
"weighing paper", or we may use either a plastic boat ("weighing boat") or another
container.

The mass of the container must be subtracted from the total mass to obtain the
mass of the material itself. This can be done by **taring** the balance. With the empty
container or weighing paper on the balance pan, the balance is set to zero by
pressing the "tare" button. When the material to be measured is placed in the
container, the mass of the material alone will be indicated.

Measure the mass of the following objects: a petri plate, a penny, a cylinder and a box of slides.

<u>Procedure 1 Using an Electronic Balance</u>

1. Make sure the measurement system of the balance is in grams (g). A "g" should be seen on the balance display. If the "g" is not present, press the "<u>mode</u>" button until the "g" appears.
2. Place the weighing boat on the balance pan.
3. Press the "tare" button. The display should read 0.00.
4. Place the object to be weighed in the weighing boat. Wait until the numbers do not change.
5. Record the mass (all of the digits on the display). **Never round your mass measurements and do not forget the units.**

<u>Results (Weights)</u>:

Petri plate ___________________________   Cylinder___________________________

Penny ___________________________   Box of Slides ___________________________

Measuring Volume

Volume is the space occupied by an object and the **liter (L)** is the basic unit. One liter is slightly larger than a quart.

Graduated Cylinders

Graduated cylinders are used to measure larger volumes. Before using, check to see what measurements it may be used for. A typical 100ml graduated cylinder is marked in 1 ml increments. Glass cylinders when containing liquids have a **meniscus**, a curve at the interface between the water and the air. (Plastic cylinders do not have a meniscus). Water attaches more strongly to glass than to plastic so the water at the top of the cylinder will "climb" the sides of the glass to form a bowl-like meniscus. To obtain an accurate measure, always position your eye, level with the meniscus and read the volume at the **lowest level**. Figure 3.

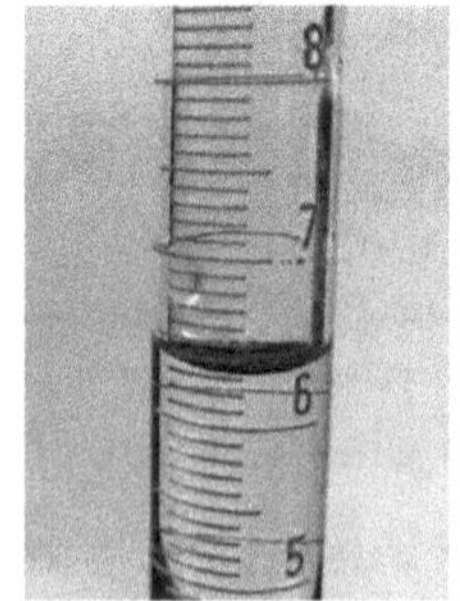

Source: Werner Williams

Figure 3 Graduated Cylinder with a Meniscus

<u>Procedure 2 Using a Graduated Cylinder</u>
1. Pour some of the red solution from its storage container into a 100ml <u>glass</u> graduated cylinder.
2. Record the volume.

   <u>Results</u>:
   Volume in glass graduated cylinder _________________________________

3. Pour **some** of the red solution from the glass graduated cylinder into a <u>plastic</u> graduated cylinder.
4. Record the volume.

   <u>Results</u>:
   Volume in plastic graduated cylinder _______________________________
   Did you see a meniscus in the plastic graduated cylinder?
   ____

5. <u>Pour the red solution from both graduated cylinders back into the storage container</u>.

<u>*Pipets*</u>
Pipets may be used to measure small volumes, usually 25ml or less. Liquids are drawn into a pipet using a pipettor. You will be using a serological pipet. Figures 4 and 5.

Guidelines:
- **Never pipet by mouth**.
- You should always keep the pipet in an almost vertical position when it contains fluid.
- To avoid contaminating the pipettor, do not allow liquid to go past the top of the pipet into the pipettor. If this happens, **let your instructor know immediately**, so it can be cleaned.
- For maximum accuracy in measuring the volume, your eye should be at the level of the top of the liquid in the pipet.

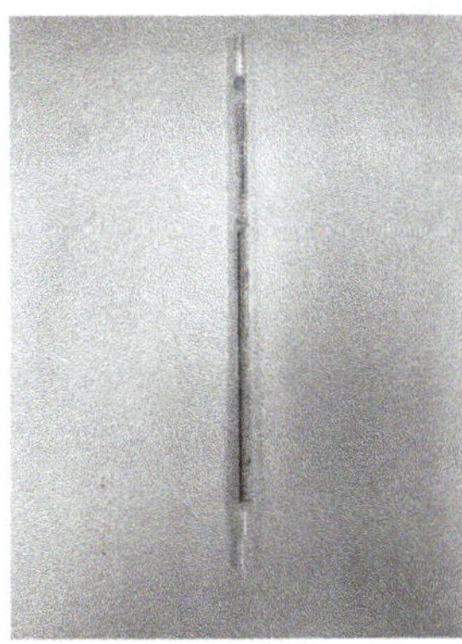
Source: Werner Williams
Figure 4 Serological Pipet

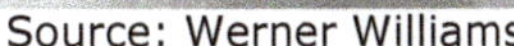
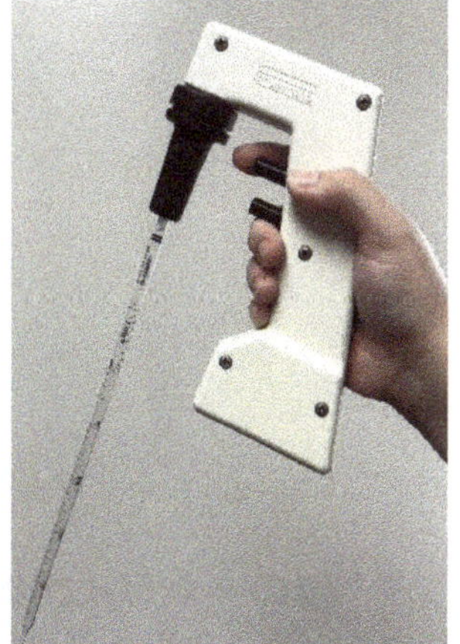
Source: Werner Williams   Source: Werner Williams
Figure 5 Pipettor with Pipet

<u>Procedure 3 Using a Serological Pipet</u>

1. <u>Insert the top end of the</u> pipet into the pipettor.
2. Insert the bottom tip of the pipet into the liquid that you want to measure.
3. Use the uppermost button to draw the liquid into the pipet; use the lower button to release the liquid. The harder you press these buttons, the faster the liquid will move into or out of the pipet. It takes practice to get used to controlling the flow of the liquid. Start slowly by pushing the buttons gently.
4. Measure the following dye solutions into an empty tube according to Table 3 below. Add each volume to the inside of the tube, allowing the liquid to flow down to the bottom. **Normally if you are taking a sample from different solutions, you would need to use a separate pipet for each solution. Today, you may use the same pipet for all three solutions.**
5. Turn off the pipettor, remove the pipet and place it back into its plastic sleeve.
6. Pour the liquid from the tube into the "waste solution" beaker and place the tube in the rack.

Table 3 Pipetting Practice Table

	Red dye vol. (ml)	Blue dye vol. (ml)	Yellow dye vol. (ml)	Total vol. (ml)
Tube	4.4	2.5	3.7	

Courtesy of Werner Williams

Micropipet
Micropipettes are used for measuring very small volumes (typically microliter [µl] amounts). Figure 6. A microliter (µl) is a millionth of a liter.
Your instructor will demonstrate the proper procedure for drawing and expelling a sample with the micropipette. The steps are listed below for you to follow.

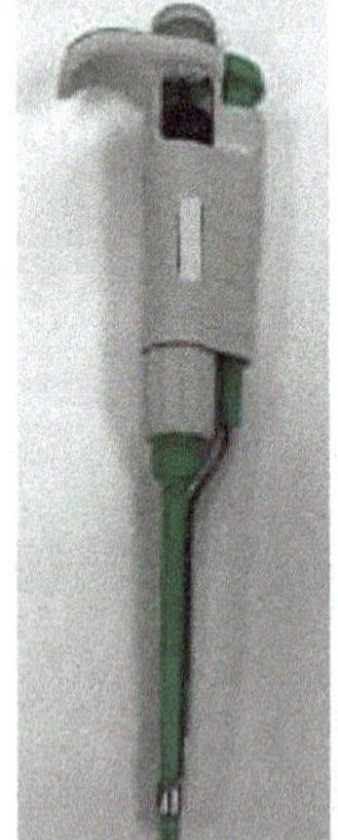

Source: Werner Williams

Figure 6 Micropipet

Source: Werner Williams

Figure 7 Micropipet Tube

Guidelines:
- Set the pipet volume only within the range for that micropipette.
- Never forget to apply a tip when using a micropipette or liquid can enter the nosecone and may damage the internal mechanism.
- Always keep the micropipette in a vertical position when there is fluid in the tip. Do not allow liquid to run back into the nose cone.
- Use your thumb to control the speed at which the plunger rises after taking up or releasing the fluid.
- If when drawing up liquid you depress the plunger to the second stop, you will be pulling up more than the amount you specified to be drawn into the tip.

Procedure 4 Using a Micropipette
1. Obtain a micropipette tube. Figure 7.
2. Add the amounts of the three solutions from Table 4 below using the same tip. **Follow steps 3-12 below to obtain and dispense the various solutions.** *Normally, we would use a different micropipette tip for each of the solutions but you may use the same tip in today's lab to obtain all three solutions.*
3. Dial the desired volume into the micropipet. Make sure you understand how to read the scale. *Hint*: By knowing the maximum volume of the micropipet (usually written on the micropipet), you can determine what each of the digits on the readout means.
4. Push the end of the pipet into the proper sized tip.
5. Hold the micropipette in one hand at a 45° angle from vertical. In this way, contaminants from your hands or the micropipette will not fall into the container you are dispensing into.
6. **Press** the plunger at the top of the micropipette to the **first** stop, and **hold** it in that position.
7. Place the tip into the solution to be pipetted.
8. Draw the fluid into the tip by *slowly* releasing the plunger.
9. Gently touch the micropipette tip to the inside wall of the tube into which you want to place the sample. (This creates a tiny surface tension effect that helps draw the fluid out of the tip).
10. Slowly press the plunger of the micropipette to the first stop, and then to the second stop to expel all of the fluid and **hold** the plunger in that position.
11. Slowly remove the pipet from the tube, *keeping the plunger depressed to avoid sucking any liquid back into the tip.*
12. When the tip is free of the container, release your thumb pressure.
13. *After dispensing all three solutions into the tube you may* eject the tip into the garbage.
14. Cap the micropipette tube and place it in the beaker labeled 'used tubes".

Table 4 Pipetting Practice Table

	Red dye vol. (µl)	Blue dye vol. (µl)	Yellow dye vol. (µl)	Total vol. (µl)
Tube	20	40	55	

Source: Werner Williams

Review:
Taking up sample: Push down to the first stop and hold→ into the liquid→ slowly release.
Releasing sample: Push down to the second stop and hold→ lift micropipette out of the container→ slowly release→ eject the tip.

What do you think would happen to the accuracy of your measurement if you depressed the plunger of a micropipette to the second stop instead of only to the first stop when taking a sample?

Flasks and beakers are used to hold liquids. Figure 8
They are not very accurate so should not be used for measuring volumes.

Source: Werner Williams
Figure 8 Beaker and Flask

Temperature
In the metric system, temperature is measured using a thermometer in degrees Celsius ($^{\circ}$C) and recorded to the nearest 0.1 degree. Figure 9.

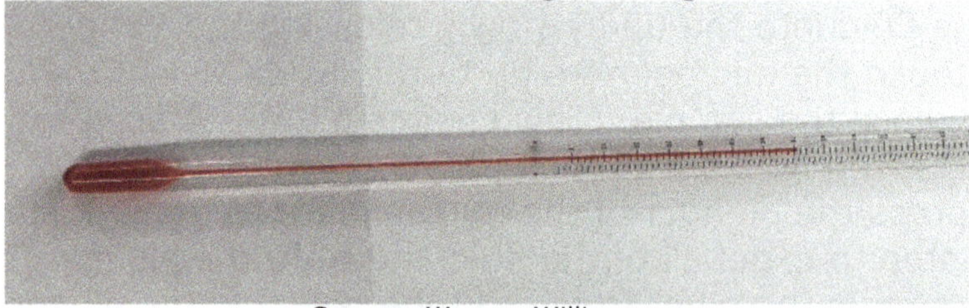

Source: Werner Williams
Figure 9 Thermometer

Procedure 5 Using a Thermometer
1. Using the thermometer that is placed in the flask of "room temperature" H_2O, record the temperature when the colored indicator solution in the thermometer no longer moves. Be sure to include the units.
2. Using the thermometer that is placed in the flask which is resting on the hot plate, record the temperature and units of the "warm" H_2O.
3. Room temperature H_2O_________________
   Warm H_2O_________________
4. Convert the temperature from $^{\circ}$C to $^{\circ}$F (Fahrenheit).

Conversion factors:

$$^\circ C = \left(\dfrac{F - 32}{9}\right) 5 \qquad\qquad ^\circ F = \left(\dfrac{^\circ C \times 9}{5}\right) + 32$$

Room temperature H_2O_______________°F   Warm H_2O_______________°F

Graphing Skills

Numerical results of an experiment are frequently presented in a graph because it is a picture of the results. Graphs show the relationship between two or more different factors. Two common types of graphs are <u>line graphs</u> and <u>bar graphs</u>. Graphs should contain at least 5 major parts:

1. **The title:** This should be a concise statement indicating what the graph is about and should be placed above the actual graph.
2. **The Independent Variable:** This is the part of the experiment (called the variable) that changes and can be controlled or manipulated by the one doing the experiment. This variable is placed on the horizontal or x-axis.
3. **The Dependent Variable**: This is the result of what happens because of the independent variable. This variable is placed on the vertical or y axis.
4. **The Scales for each Variable**: In constructing a graph, the data points should be plotted as a scale. The scale should have all the data points, should make good use of the space available and should have consistent increments. It should also be easy to manage. Data points with increments of 4 or 10 are easy to manage, but increments of 2.37 or 4.93 are not.
5. **The Legend**: This should be a short description of the graph's data. It should be placed directly under or next to the graph.

<u>Line Graphs</u>

A <u>line graph</u> shows the best relationship between two variables. It has one or more lines connecting a series of points. Figure 10. Along the horizontal or x-axis, you will find the variable that has been manipulated. Along the vertical or y-axis, you will find the resulting variable.

The scale should be chosen to make the graph as large as possible within the limits of the paper. If the scale unit is too large, your graph will be cramped into a small area and will be hard to read and interpret. If the scale is too small, the graph will run off the paper.

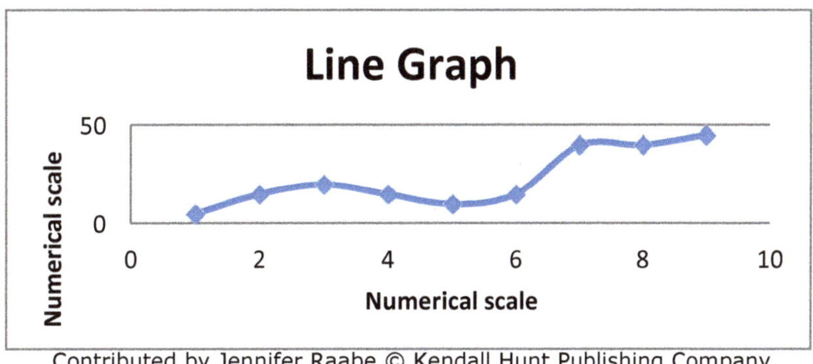

Contributed by Jennifer Raabe © Kendall Hunt Publishing Company

Figure 10

Use the line graph in Figure 11 to answer questions 1-6 below.

<u>Questions</u>
1. Which plant grew the tallest? ___________________________________
2. How many plants grew to be at least 6 cm tall? _________________
3. Which plant grew the fastest in the first five days? _______________
4. Which line represents plant #2? _________________________________
5. After 10 days, how much had plant #3 grown? _________________
6. How long did it take for plant #1 to grow 6 cm? _________________

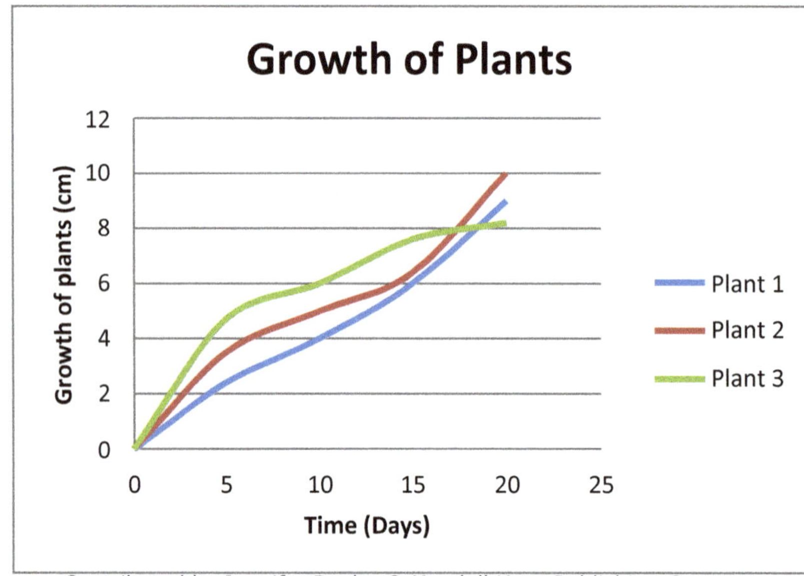

Contributed by Jennifer Raabe © Kendall Hunt Publishing Company

Figure 11

Bar Graph
A <u>bar graph</u> is another way of showing the relationship between variables. Instead of constructing a line, a bar graph uses columns to display the data. Figure 12

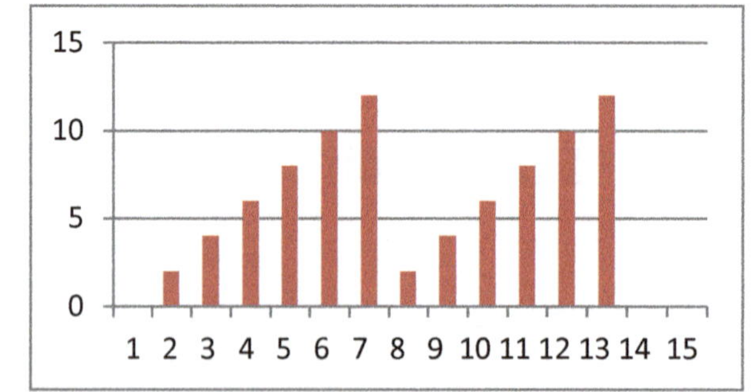

Contributed by Jennifer Raabe © Kendall Hunt Publishing Company
Figure 12

Use the bar graph in Figure 13 below to answer questions 1-5.

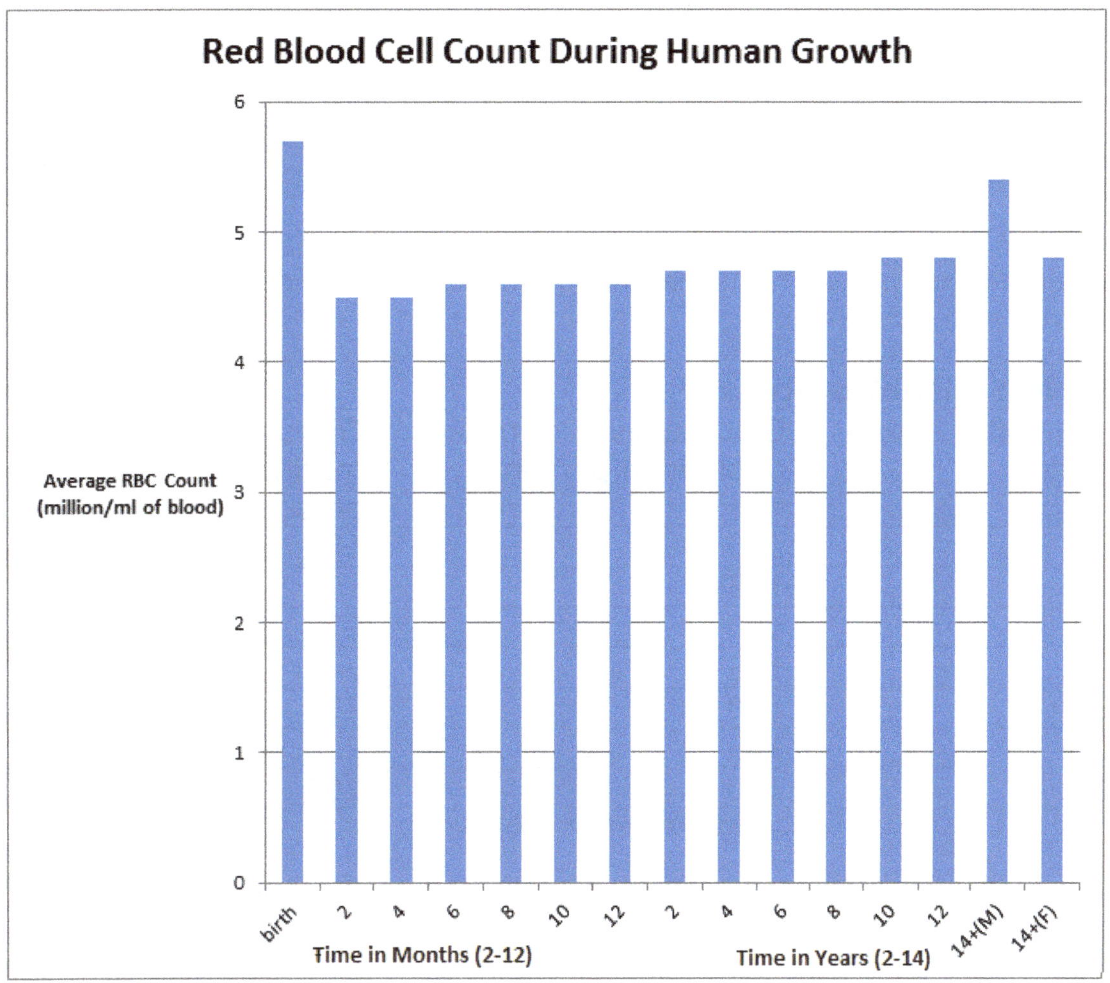

Contributed by Jennifer Raabe © Kendall Hunt Publishing Company

Figure 13 Red Blood Cell Count From Birth to Age 14

<u>Questions</u>

1. At birth, what is the average number of red blood cells (rbc's) per ml of blood? ____

2. What appears to happen to the number of rbc's between birth and 2 months? ____

3. What happens to the number of rbc's between the age of 6 and 8 years? ____

4. Between what ages is a human likely to have 4.6 million rbc's? ____

5. After 14 years of age, do males (M) or females (F) have a higher red blood cell count? ____

Procedure for constructing a line graph

Diabetes is a disease that affects the insulin-producing cells of the pancreas. If the pancreas does not produce enough insulin, the amount of glucose in the blood will remain high. A blood glucose level above 140 for an extended period of time is not considered normal. It may be a sign of diabetes. If diabetes is not brought under control, it will lead to severe complications and even death.

Use the data in the table below (Table 5) to construct a line graph. Remember to title it, label the axes properly when setting up the scale, and write a legend for the graph.

Table 5 Diabetes Study

Time after eating (hrs.)	Glucose level in ml/liter of blood in person A	Glucose level in ml/liter of blood in person B
0.5	170	180
1	155	195
1.5	140	230
2	135	245
2.5	140	235
3	135	225
4	130	200

Source: Werner Williams

Answer the following questions based on the graph on the previous page and from Table 5.

1. What is the independent variable? ____

2. Why is that the independent variable? ____________________________________
    ____

3. What is the dependent variable? ____

4. Why is that the dependent variable? _____________________________________
    ____

5. Which if any of the individuals has diabetes? Justify your answer.
    ____
    ____
    ____
    ____

6. If the time period were extended to 6 hours, what would be the expected
    blood sugar level for person B? ____

7. What would be a probable blood sugar level for person B at 3.5 hours?
    ____

Lab 2: Use of the Scientific Method

Process of Science
- Science is based on repeatable observations and hypotheses that are testable.
- It is limited to what can be observed and measured.
- Science looks for natural causes for natural events.
- The generally accepted method that scientists use to determine "the facts" is called the scientific method.

Scientific Method
Scientists often use a series of steps to efficiently investigate a question or phenomenon in nature. Study the steps below of the scientific method.
1. Observation – Observe some aspect of nature.
2. Question – Frame a question to identify a problem or to explain a phenomenon.
3. Hypothesis – Collect background information and make an educated guess about a possible answer or solution to the problem.
4. Prediction – Make a statement of what you expect to find based on the hypothesis.
5. Test your prediction by performing experiments. (Collect data.)
6. Conclusion – Analyze the results to determine if the hypothesis is correct.

Procedure 1
For each sentence below, enter the letter of the correct step of the scientific method.

a. Testing of a hypothesis by experiment or observation
b. States an hypothesis
c. States results (facts only)

_________1. Bean plants may grow faster in the presence of a fertilizer.

_________2. In an experiment, 68 out of 85 flies were attracted to honey.

_________3. A student in a microbiology class grew bacteria on media plates that contained different brands of mouthwash.

_________4. A group of genetics students designed and distributed a survey to the student body concerning the presence of freckles on faces.

________5. Geraniums grown under red light will grow faster than under green
      light.

________6. An experiment showed that aphid insect species on tomato plants are
      more resistant to a particular insecticide than caterpillar insect species.

Forming a Hypothesis and Making a Prediction

A **hypothesis** is an educated guess about a possible answer or solution to a
problem. The hypothesis should be testable and falsifiable. Logic and reasoning are
invaluable tools in science to help formulate hypotheses.

Procedure 2
Circle **YES** if you think the hypothesis is good (i.e., testable) as written, and **NO** if
you think it is not. If you choose **NO**, indicate how you would change the
hypothesis to make it a good one.

1. Students who own laptops have higher GPAs. **YES** or **NO**
2. Murders occur more often during a full moon. **YES** or **NO**
3. Cats are happier when you pet them. **YES** or **NO**
4. Sea level will be higher in 100 years than it is today. **YES** or **NO**

____

____

____

____

The process of making a hypothesis uses both **inductive and deductive
reasoning**.
   Inductive reasoning:
   - Collect background information related to your observations.
   - Make a general statement from specific observations before making
     the hypothesis.
   - This can be used to develop a testable hypothesis.
   Deductive reasoning:
   - Move from general observations to specific results.
   - Use the "if-then" reasoning format.
   - Make a **prediction** that can be used to test the hypothesis.

Procedure 3:

Study the observations and questions given below in Table 1. Complete the table giving possible hypotheses and predictions. Remember to put the predictions in the "if-then" format.

Table 1

Observation	Question	Hypothesis	Prediction
1. Heart rate is steady when we are at rest sitting.	If we are standing or holding our breath, will our heart rate speed up?		
2. People can hold their breath for a short period of time.	Can we hold our breath longer after we exercise?		
3. Tall people have long arms.	Do tall people have longer arms than shorter people?		

Source: Werner Williams

Experimentation

An experiment is designed to isolate a factor or **variable** that you are interested in testing. Experiments should be well designed so that others should be able to conduct the same experiment and get the same results. In most experiments, subjects, individuals, or specimens to be tested are divided into two groups: an **experimental group** that is treated with the **experimental variable** and a **control group** that is not subjected to the experimental variable. An "if-then" prediction is often useful in designing an experiment.

There are at least two categories of variables. The **independent variable** is the one factor to be tested. The **dependent variable** is the result that is measured or observed at the end of the experiment.

Collection and Interpretation of Data

Tables and Graphs

Collected data from a laboratory experiment may be presented in a table form. Tables should always include a descriptive title with column and row titles, including units of measure. However, numerical data is often presented in graph form, either a bar graph or a line graph. In designing an experiment the **dependent variable** is expected to change in response to changes in the **independent variable**. In other words, the independent variable will influence the outcome of the dependent variable. On a graph, the dependent variable is placed on the **y** axis and the independent variable is placed on the **x** axis. Additional rules to consider in graphing data include: properly label each axis, specify measuring units on each axis, use appropriate intervals for the range of data so that most of the area of the graph is utilized, and always include a title that very briefly describes the experimental conditions.

In the example below, the control group plants and the experimental group plants (fertilized plants) were treated the same (same amount of light, same water etc.), but the fertilized plants were the only ones exposed to fertilizer, the experimental variable. The results are shown below and on the next page.

Collected Data in Table Form:

Table 2

Fertilizer Experiment: Height Gain (cm) of Plants Over a Four Week Period					
Specimen	**Initial Height (cm) start**	**Week 1**	**Week 2**	**Week 3**	**Week 4**
Control plant	5.0	6.6	7.3	9.9	11.1
Fertilized plant	5.1	8.9	12.2	15.5	19.4

Source: Werner Williams

© Alistair Scott/Shutterstock.com

Bar Graph

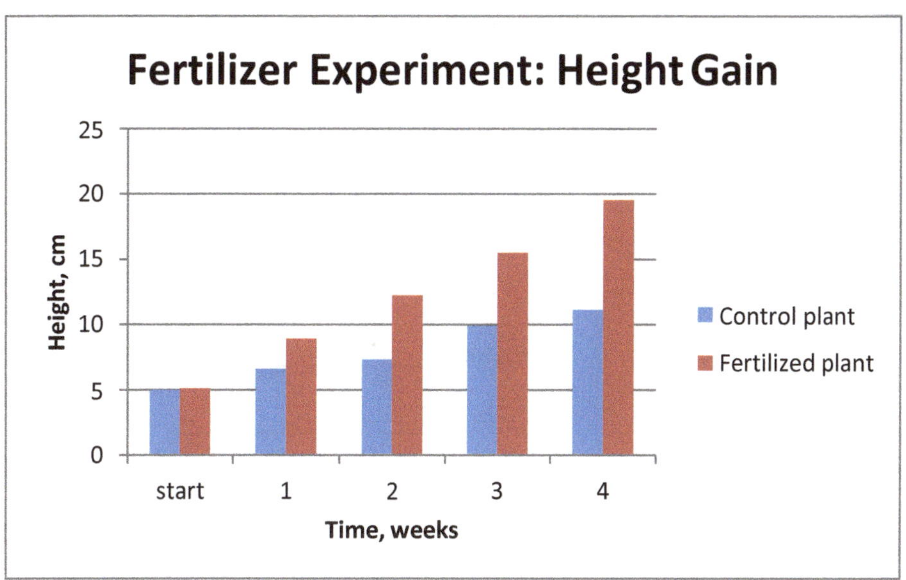

Contributed by Jennifer Raabe © Kendall Hunt Publishing Company

Line Graph

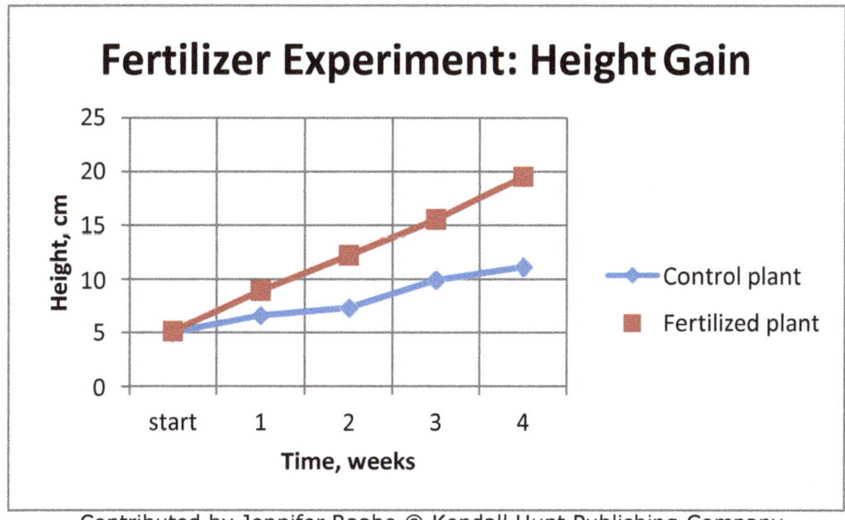

Contributed by Jennifer Raabe © Kendall Hunt Publishing Company

Simple Statistics

Simple statistical analysis is often used to help interpret the results of an experiment with a large amount of data. The most commonly used measures are the mean, the mode, the median and the range. The **mean** is what you commonly think of as the average of a set of numbers. Add all numbers in the set and divide by the number of samples in the set to determine the mean. The **mode** is the most frequently occurring number in a set. The **median** is the middle number of a set when all numbers are arranged in either ascending of descending order. The **range** is the <u>difference </u>(one number) between the largest number and the smallest number in a set.

<u>Procedure 4</u>
Use the data in the Table 3 below to determine the mode, mean, median, and range for the student scores.

Table 3

Student	1	2	3	4	5	6	7	8	9	10	11	12	13	14	15
Quiz score	91	82	80	76	71	43	84	62	94	82	35	55	66	82	74

Source: Werner Williams

Mode:_____________________   Median:__________________

Mean: ___________________   Range: __________________

<u>Procedure 5 Rats in a Maze</u>

Study the experimental data presented in Table 4 to answer the questions that follow.

© Fer Gregory/Shutterstock.com

Table 4

The Effect of Ingested Alcohol Upon Maze Running Performance in Rats		
Rat ID	**Alcohol Ingested (ml/kg body mass)**	**Maze Run Time (minutes)**
1	0	0.21
2	0	0.25
3	0	0.20
4	0.1	0.34
5	0.1	0.29
6	0.1	0.31
7	0.3	1.16
8	0.3	1.42
9	0.3	1.33
10	0.5	2.74
11	0.5	3.38
12	0.5	3.05
13	0.7	5.60
14	0.7	6.18
15	0.7	6.40
16	0.9	9.20
17	0.9	8.83
18	0.9	10.35

Source: Werner Williams

1. What hypothesis was being tested by this experiment?

2. What is the independent variable for this experiment?

3. What is the dependent variable for this experiment?

4. Which rats served as the controls for this experiment? How do you know?

<u>Performing an Experiment and Collecting Data</u>

The human immunodeficiency virus (HIV), which causes the disease AIDS, is sexually transmitted. This activity will simulate how multiple sexual encounters can lead to the increasing spread of such a disease through a population.

Human immunodeficiency virus (HIV)

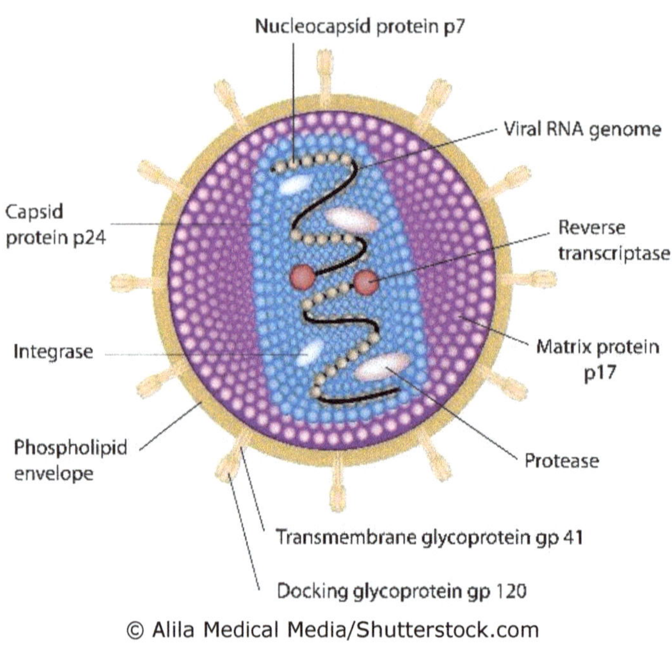

© Alila Medical Media/Shutterstock.com

<u>Procedure 6 Spread of the HIV</u>
See separate handout from instructor.

<u>Procedure 7</u>

Choose, along with the members in your group, one prediction/hypothesis from Table 5 below to conduct as an experiment. Consider the three paragraphs and the procedures/instructions following the table that you would use before choosing which hypothesis to test.

Table 5

Observation	Question	Hypothesis	Prediction
1. Heart rate is steady when we are at rest sitting.	If we are standing or holding our breath, will our heart rate speed up?	Our heart rate will stay the same whether we are sitting, standing or holding our breath.	If the hypothesis is correct, then our heart rate should stay the same, whether we are sitting, standing, or holding our breath.
2. We can hold our breath for a short period of time.	Can we hold our breath longer after we exercise?	We can hold our breath longer immediately after they exercise.	If the hypothesis is correct, then after exercise, we should be able to hold our breath for a longer period of time.
3. Tall people have long arms.	Do tall people have longer arms than shorter people?	The taller the person, the longer his or her arm length becomes.	If the hypothesis is correct, then taller people should have longer arm lengths than shorter people.

Source: Werner Williams

To test the first prediction, the test subjects should have their pulse checked at their wrist (which tells us their heart rate) while sitting, have it checked while sitting and holding their breath, and then have it checked while they are standing.

To test the second prediction, the test subjects should be tested to see how long they can hold their breath at rest while sitting. Then they should walk down and up the stairs **twice** very quickly. Immediately after entering the lab, have them sit down and see how long they can hold their breath.

To test the third prediction, the test subject's height should be measured using a meter stick and the length of both of their arms should be measured with a tape measure.

<u>Instructions for Procedure 7</u>

Hypothesis #1 (Pulse Rate)
1. Invite 10 students to be your test subjects. They should be as varied as possible. Include equal numbers of males and females. Include yourselves.
2. Have the subjects sit quietly, find their pulse on their wrist and using a timer count the number of beats for 1 minute. **Do not take your own pulse.**
3. Repeat step 2 for each subject while the person stands.
4. Repeat step 2 for each subject but have the person sit while holding their breath. (If you can't hold your breath for 1 minute, do it for 30 seconds and multiply by 2).
5. Record the results.

Hypothesis #2 (Holding Breath)
1. Invite 10 students to be your test subjects. They should be as varied as possible. Include equal numbers of males and females. Include yourselves.
2. Obtain a timer. While the subjects are sitting, have them hold their breath as long as they can while timing them.
3. Have the subjects walk down and up the stairs **twice** very quickly, and immediately on entering the lab, have them sit and hold their breath while timing them.
4. Record the results.

Hypothesis #3 (Height and Arm Length)
1. Invite 10 students to be your test subjects. They should be as varied as possible. Include equal numbers of males and females. Include yourselves.
2. Have your subjects stand and using a meter stick, measure their height in cm.
3. Using a tape measure, measure the length of both arms in cm, add the lengths and divide by two to get that person's average arm length.
4. Record the results.

Hypothesis to be tested:

____

____

Before you begin, notify your instructor which experiment you will be conducting.

Results:

Table 6 Pulse Rate Results

Test Subject	Heart Rate Sitting	Heart Rate Standing	Heart Rate While Holding Breath
1			
2			
3			
4			
5			
6			
7			
8			
9			
10			

Source: Werner Williams

Table 7 Holding Breath Results

Test Subjects	Holding Breath at Rest	Holding Breath after Exercise
1		
2		
3		
4		
5		
6		
7		
8		
9		
10		

Source: Werner Williams

Table 8 Height and Arm Length

Test Subject	Height (cm)	Average Arm Length (cm)
1		
2		
3		
4		
5		
6		
7		
8		
9		
10		

Source: Werner Williams

<u>Rearrange the data from Table 8 into Table 9, placing the data from the shortest person to the tallest person.</u>

Table 9 Height and Arm Length from Shortest Test Subject to Tallest

Test Subject	Height (cm)	Average Arm Length (cm)
1(Shortest)		
2		
3		
4		
5		
6		
7		
8		
9		
10(Tallest)		

Source: Werner Williams

Graph the results of your experimentation:

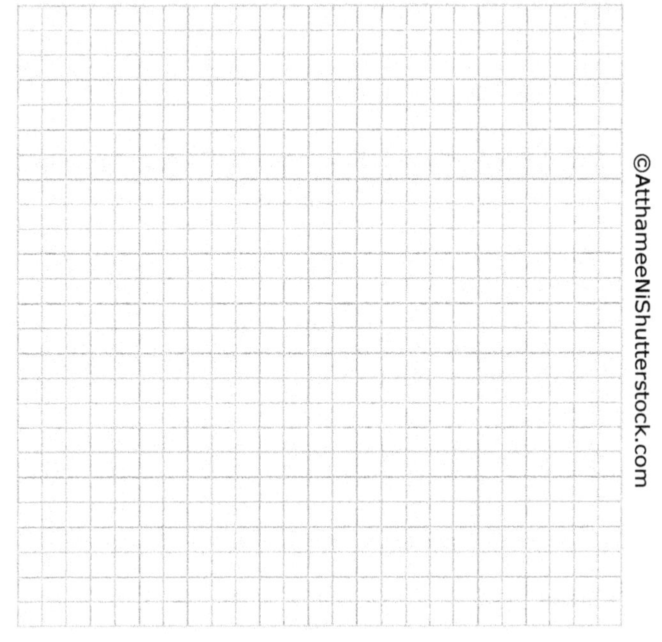

Average Heartbeats (beats/min.)
Sitting
Standing
Holding Breath
©AtthameeNiShutterstock.com

Average Time of Breath Holding (seconds)

After Rest   After Exercise

Arm Length (cm)

Height (cm)

<u>**Stating a Conclusion**</u>

Discuss the results from Procedure 7 with your other group members.

Based on your hypothesis and resulting data from experimentation, write a conclusion for Procedure 7. Support your conclusion with facts collected during the experiment.

Lab 3: Atoms and Molecules

The **periodic table** (Figure 1) lists information about the elements that have been discovered to this date. Each element is given a box, identifying its symbol and the number of its electrons, protons and neutrons in its atoms. The number at the top is the **atomic number**, which is the number of protons or electrons in that element's atom. The **mass number** is the number at the bottom and is the total number of protons and neutrons.

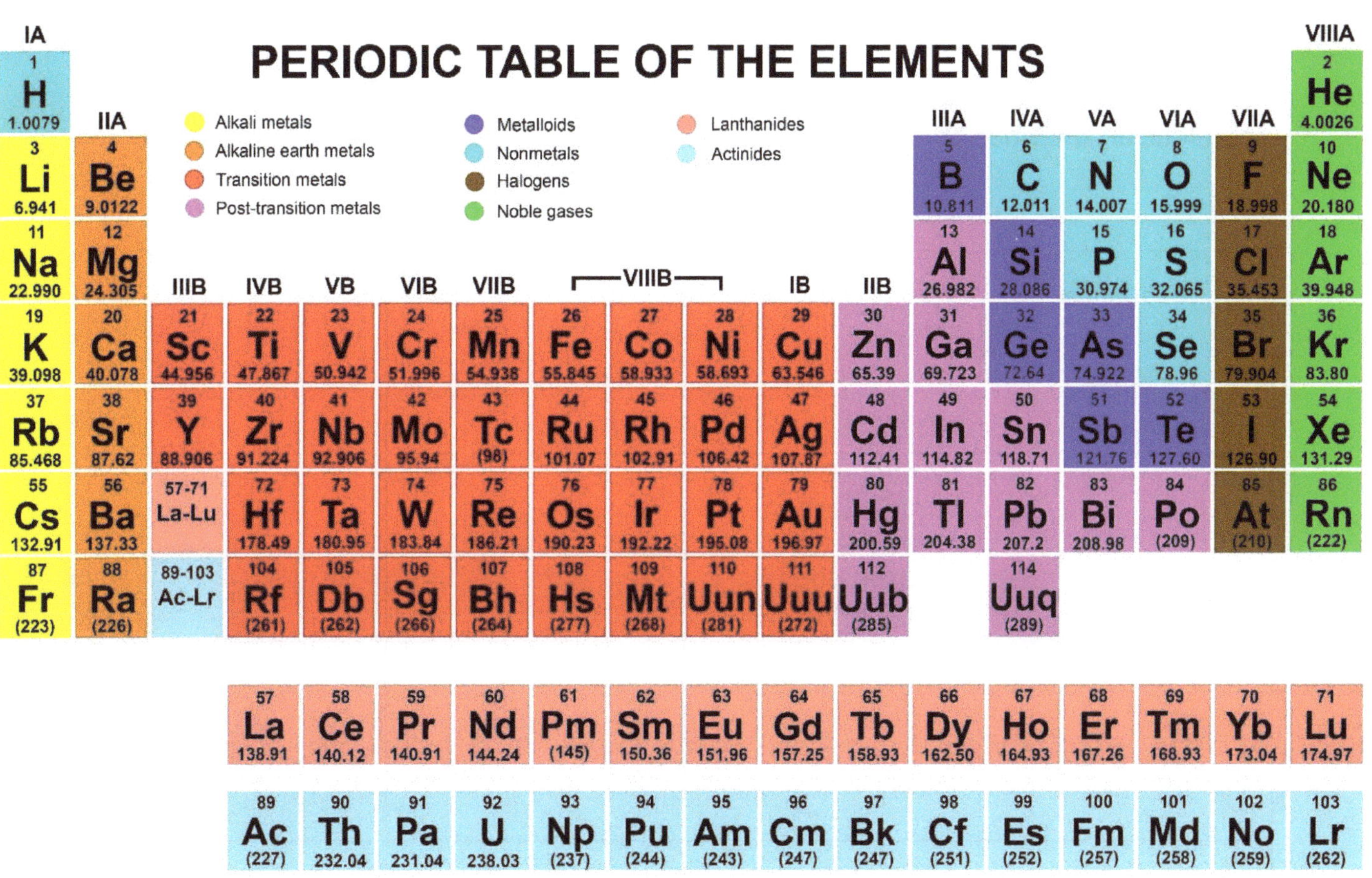

Figure 1 The Periodic Table

Figure 2 below illustrates information about a carbon atom.

© Andrei_M/Shutterstock.com

Figure 2 Information about Carbon

<u>Procedure 1</u>

For the elements with the atomic numbers 1-20, list their name, symbol, atomic # and electron distribution in Table 1 below. Carbon's is given as an example below.

Table 1

	Symbol- C Name- Carbon Atomic # - 6 Electron shell distribution-2,4		

Source: Werner Williams

<u>**Ionic and Covalent Bonds**</u>

Most atoms can be combined in certain ways to form larger structures called molecules or compounds. Some atoms have a strong attraction for electrons and can steal electrons from other atoms. When an atom gains or loses electrons, it becomes an **ion**. If an atom gains electrons, it becomes negatively charged (-) and if it loses electrons, it becomes positively charged (+). If ions have the same charge, (both + or both -), they will repel each other, while ions that have opposite charges (+ and -) will attract each other and form an **ionic bond**.

<u>Procedure 2</u>

Use the periodic table to determine the number of protons and electrons that are found in the following atoms, ions and molecules. When you determine the amount of protons or electrons in a particular ion, the number of plus signs (+, +2 and so on) determines the number of electrons that were lost and the number of minus signs (-, -2 and so on) determines the number of electrons that were gained.

1. Protons in a sodium atom (Na)__________Electrons in a sodium atom (Na) ____________

2. Protons in a sodium ion (Na^+)__________Electrons in a sodium ion (Na^+) ____________

3. Protons in a chlorine atom (Cl)__________Electrons in a chlorine atom (Cl) ____________

4. Protons in a chloride ion (Cl^-)__________Electrons in a chloride ion (Cl^-) ____________

5. Protons in a sodium chloride molecule (NaCl) _______
    Electrons in a sodium chloride molecule (NaCl) _____

6. Protons in a calcium atom (Ca)__________Electrons in a calcium atom (Ca) ____________

7. Protons in a calcium ion (Ca^{+2})__________Electrons in a calcium ion (Ca^{+2}) ____________

8. Protons in an oxygen atom (O)__________Electrons in an oxygen atom (O) ____________

9. Protons in an oxide ion (O^{-2})__________Electrons in a oxide ion (O^{-2}) ____________

10. Protons in a calcium oxide molecule (CaO)_____Electrons in a calcium oxide molecule (CaO) _ _

11. Protons in a magnesium atom (Mg)______Electrons in a magnesium atom (Mg) __________

12. Protons in a magnesium ion (Mg^{+2})______Electrons in a magnesium ion (Mg^{+2}) __________

13. Protons in a chlorine atom (Cl)_________Electrons in a chlorine atom (Cl) ______________

14. Protons in a chloride ion (Cl⁻)__________Electrons in a chloride ion (Cl⁻) ______________

15. Protons in a magnesium chloride molecule ($MgCl_2$) __________

   Electrons in a magnesium chloride molecule ($MgCl_2$) __________

Procedure 3

Consider Table 2 below. Some positive and negative ions are listed. List at least 5 compounds formed by combining <u>any particular</u> + ion with <u>any particular</u> – ion. **(Do not use either of the 2 examples given below)**.Write the symbol for the positive ion first and then the negative ion. You may also use a subscript to indicate the number of ions needed to balance the charge. For example: Na_2SO_4, $Pb(NO_3)_2$

Table 2

Positive Ions	**Negative Ions**		**Formed Compounds**
H⁺	NO_3^{-1}	(nitrate)	
Na⁺	O^{-2}	(oxide)	
Mg^{+2}	SO_4^{-2}	(sulfate)	
Ca^{+2}	Cl^{-1}	(chloride)	
Ag^{+1}	OH^{-1}	(hydroxide)	
Pb^{+2}	HCO_3^{-1}	(bicarbonate)	

Source: Werner Williams

Covalent bonds are a second type of bond that holds atoms together to form molecules. Electrons are shared by two or more atoms. We can diagram a molecule by allowing a line to represent a single covalent bond.

Figure 3 shows a single carbon atom (C) sharing electrons with four different hydrogen atoms (H), and each of the four hydrogen atoms is sharing a pair of electrons with the same C atom. That molecule is called methane.

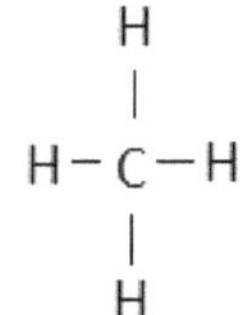

H
|
H − C − H
|
H

Source: Werner Williams

Figure 3

Sometimes two atoms may share more than one pair of electrons, creating a double bond. For example, in a carbon dioxide molecule (CO_2), a carbon atom is sharing 2 pairs of electrons with one oxygen atom and 2 pairs of electrons with another oxygen atom ($O=C=O$).

Table 3 below lists a few atoms and the number of electrons they usually share.

Name of Atom	Symbol of element	Number of bonds	Bonding capacity
Carbon	C	4	=C= −C− =C
Nitrogen	N	3	−N= ∕N∖ N≡
Oxygen	O	2	O= O−
Hydrogen	H	1	H−

Source: Werner Williams

The Lewis dot structure displays the number of electrons in the valence shells of each atom in a molecule. For example, C has 4 valence electrons while H has 1. These valence electrons in a Lewis dot structure are arranged in such a way that each atom has a full valence shell by sharing some or all of their valence shell electrons. Hydrogen's valence shell is its first shell and only requires 2 electrons to fill, while carbon's valence shell is its second shell and requires eight electrons to fill. So each H should have 2 valence electrons around it and each C should have 8. See the Lewis dot structure of methane in Figure 4.

H
••
H : C : H
••
H

Contributed by Kim Hakala. © Kendall Hunt Publishing Company.

Figure 4

<u>Procedure 4</u>

Construct models of each of the molecules in the table below by using the wooden balls (Black-C; Yellow or White-H; Red-O; Blue-N), sticks and springs (bonds). Diagram the structural formula and Lewis dot structure of each molecule you build.

Molecular formula	Structural formula	# of bonds	Lewis dot structure	Total valence electrons
1. H_2	H-H	1	H : H	2
2. O_2				
3. H_2O				
4. CO_2				
5. NH_3				
6. CH_4				
7. C_2H_4				
8. $C_3H_6O_3$				

Source: Werner Williams

<u>**Reactions**</u>

A chemical reaction takes place when different molecules react with each other so that chemical bonds are broken and as new bonds are formed, new molecular combinations are created. In many cases, there is evidence that a reaction has taken place such as bubbles being released, a color change or the formation of a solid that settles on the bottom of the container.

We can use a chemical equation to show a chemical reaction, with the starting materials (**the substrates or reactants**) on the left and the new substances called **the products** on the right. The arrow indicates the direction of the reaction.

For example, hydrogen atoms react with oxygen to form water:

$$2 \, H_2 \; + \; O_2 \; \rightarrow \; 2 \, H_2O$$

Count the number of atoms on each side of the reaction arrow.

1. How many H atoms are on the left? ___________________________________

2. How many H atoms are on the right? _________________________________

3. How many O atoms are on the left? __________________________________

4. How many O atoms are on the right? _________________________________

When the number of each kind of atom on each side of the equation are the same, the equation is said to be **balanced**. All reactions should be balanced.

<u>Procedure 5</u>

Use extreme caution when mixing all chemicals.

1. Consider the following equation:

$$NaCl + AgNO_3 \rightarrow NaNO_3 \; + \; AgCl$$

a. Complete the equation by writing out the name of the products:

Sodium chloride + Silver nitrate →______________________+ _____________________

b. Add a small amount of NaCl solution to a test tube. Add a few drops of $AgNO_3$.

c. What change did you observe that would indicate that a reaction took place?____

2. Consider the following equation:

$$2\ NaI + Pb(NO_3)_2 \rightarrow PbI_2 + 2\ NaNO_3$$

a. Complete the equation by writing out the name of the products:

Sodium iodide + lead nitrate → ______________________________+______________________

b. Add a small amount of NaI solution to a test tube. Add a few drops of $Pb(NO_3)_2$.
c. What change did you observe that would indicate that a reaction took place?____

3. Consider the following equation:

$$HCl + NaHCO_3 \rightarrow CO_2 + \underline{\hspace{3cm}} + \underline{\hspace{3cm}}$$

a. Complete the equation by writing in the chemical symbols of the products.

b. Add a small amount of HCl solution to a test tube. Add a few drops of $NaHCO_3$.
c. What change did you observe that would indicate that a reaction took place?____

Lab 4: Solutions and pH

Solutions

Living organisms contain chemicals that are in solutions. A solution is a mixture of a solute, a dissolved substance such as salt or sugar and a solvent, the dissolving material such as water. The concentration of a solute is frequently expressed as a percentage of the total solution. For example, a 7% solution (weight/volume) of sugar is made by dissolving 7g of sugar in 100ml of water or 70 g of sugar in 1000ml (1L) of water.

Molarity

The most common way to measure concentration is called molarity. A mole of a substance contains 6.02×10^{23} (Avogadro's number) molecules, atoms or ions of that substance.

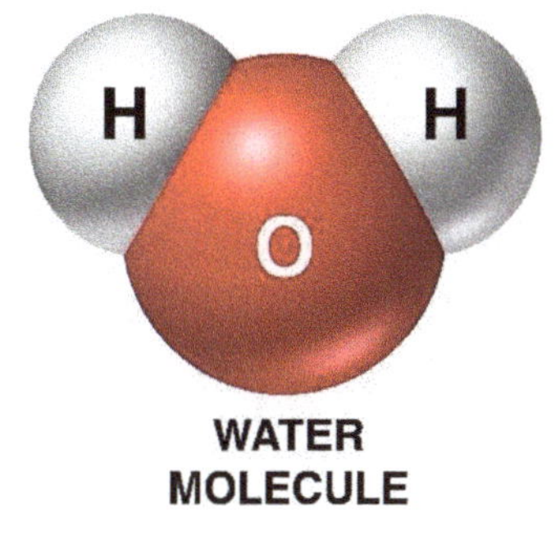

© Lightspring/Shutterstock.com

$$6.02 \times 10^{23}$$

© SedovaY/Shutterstock.com

1 molecule of water   Avogadro's number of molecules   1 mole H_2O = 18.0g
(1 mole)

1 mole of water and 1 mole of salt (NaCl) contain the same number of molecules. However, they weigh different amounts because a molecule of water and a molecule of salt have different weights. This weight or mass called the molecular weight is determined by adding the atomic weights of the elements contained in it.

Atomic Weights of the First 20 Elements of the Periodic Table

Elements	Symbol	Atomic Wt.		Elements	Symbol	Atomic Wt.
1.Hydrogen	H	1.0g		11.Sodium	Na	23.0g
2.Helium	He	4.0g		12.Magnesium	Mg	24.3g
3.Lithium	Li	7.0g		13.Aluminum	Al	27.0g
4.Beryllium	Be	9.0g		14.Silicon	Si	28.1g
5.Boron	B	10.8g		15.Phosphorus	P	31.0g
6.Carbon	C	12.0g		16.Sulfur	S	32.1g
7.Nitrogen	N	14.0g		17.Chlorine	Cl	35.5g
8.Oxygen	O	16.0g		18.Argon	Ar	40.0g
9.Fluorine	F	19.0g		19.Potassium	K	39.1g
10.Neon	Ne	20.2g		20.Calcium	Ca	40.1g

Source: Werner Williams

The molecular weight of H_2O is =18g/mole [2 times the atomic weight of hydrogen=2x1.0=2.0g; plus the atomic weight of oxygen=16.0g; 2+16=18.0g]. One mole of water weighs 18.0g.

The molecular weight of NaCl is =58.5g/mole [the atomic weight of sodium=23.0g; plus the atomic weight of chlorine=35.5g; 23+35.5=58.5g]. One mole of NaCl weighs 58.5g.

© pedphoto36pm/Shutterstock.com

© Aksenenko Olga/Shutterstock.com

© Singkham/Shutterstock.com

Sulfur
1 mole= 32g

Copper
1 mole= 64g

Water
1 mole= 18g

All of the above substances (sulfur, copper, water) contain 6.02×10^{23} particles of that particular substance, which is equal to <u>1 mole.</u>

A <u>molar (M)</u> solution contains 1 mole of a solute in a liter of solution. To prepare a liter of a 1M solution of NaCl, you would add 58.5g of NaCl to a 1L (1000ml) volumetric flask and add water to the 1000ml mark. The total volume of the solvent and solute must equal 1000ml. Remember that a molar solution is defined as 1 mole of a substance per liter of <u>solution</u>, not per liter of water (solvent). If you added 1000ml of water to 58.5g of NaCl, you would have more than a liter of solution. That is called a <u>molal</u> solution.

To make a 1M solution of magnesium fluoride (MgF_2):

© Kei Shooting/Shutterstock.com

© focal point/Shutterstock.com

Add 62.3g of MgF_2 to a container   +   add water to total
1L (1000ml)

Calculations

1. How would you make 1 liter of a 1M solution of $CaCl_2$? (1 mole of $CaCl_2$ has a molecular weight of 111g).

2. How would you make a liter of a 0.4M solution of NaCl?
(To determine how many grams of NaCl you would use, simply multiply the molecular weight of NaCl by the molarity which is .4).

3. How would you prepare 500ml of a .6M solution of NaCl?
(To determine the number of grams of solute, use the following formula:
Grams of solute= molecular weight in grams x volume [in liters {500ml = .5L}] x molarity)

4. How would you prepare 250ml of a 3M solution of NaCl?

Procedure 1 Preparing a Solution

Your instructor will ask your group to prepare 100ml of a NaCl solution. The molarity assigned to your group will be selected from a range of 0.2 to 1.4M.

Molarity of NaCl for your group________Grams of NaCl required____________
Prepare your solution using a 100ml volumetric flask. Bring it to your instructor and explain how you made it.

Dilutions
A volume to volume dilution is the ratio of the solute to the final volume of the diluted solution. For example, to make a 1:10 dilution of a 1M $AgNO_3$ solution, you would mix "one part" of the 1M solution with "nine parts" of water (solvent) for a total of "10 parts." So a 1:10 dilution means 1 part+9 parts= 10 parts. If you needed 10 ml of a 1:10 dilution, you would mix 1 ml of $AgNO_3$ + 9 ml of water. If you needed 100ml of a 1:10 dilution, then you would mix 10 ml of $AgNO_3$ + 90 ml of water. The final concentration of both would be 0.1M.

It is very efficient to prepare concentrated <u>stock solutions</u> of chemicals which can then be diluted with water to make new solutions. There is an equation that is useful in determining how much of the stock solution to dilute with water to make a new solution. The equation is as follows:

$V_iC_i=V_fC_f$ V_i=initial volume; V_f= final volume

C_i= initial concentration; C_f= final concentration

Calculations
1. How much of a .5M solution should we use to make 600ml of a .4M solution?
   Initial volume(x); initial conc.=.5M; final volume=600ml; final conc.=.4M

$V_iC_i=V_fC_f$ $x(.5M)=600ml(.4M)$ $x=600(.4)/.5=480ml$

Therefore, we need 480ml of a .5M solution and add 120ml of water to get a final volume of 600ml at a conc. of .4M.

2. How many ml of water would you need to add to 100ml of 2.0M HCL to prepare a final solution of .25M HCL?

Initial volume=100ml; initial conc.=2.0M; final volume(x); final conc.=.25M

$V_iC_i=V_fC_f$ $100ml(2.0M)=x(.25M)$ $x=100(2.0)/.25=800ml$

We need to add 800ml of H_2O to 100ml of a 2.0M HCl to get a final conc. of .25M.

3. How many ml of water would you need to add to 250ml of .6M KOH to prepare a final solution of .45M KOH?

4. How much of a 3.6M solution should we use to make 200ml of a 2.8M
    solution?

pH

Some molecules that are dissolved in water may dissociate (separate or break
apart) into ions. Water itself may dissociate into H^+ (hydrogen ions) and OH^-
(hydroxide ions).

$$H_2O \rightarrow H^+ \quad + \quad OH^-$$

In any given volume of water, a small but constant number of H_2O molecules are
dissociated. In pure water, the number of H^+ is equal to the number of OH^- since
one cannot be formed without the other being formed.

The pH of a solution is the measure of the concentration (conc.) of the H^+ in that
solution. A solution of pH 7 is **neutral**, is at the midpoint of the pH scale and it
contains an equal amount of H^+ and OH^-. Solutions with a pH value lower than 7 are
said to be **acidic**. In an acidic solution, the number of H^+ exceeds the number of
OH^-. Solutions with a pH above 7 are **basic** or **alkaline** and the number of OH^-
exceeds the number of H^+. Expressed another way, the more acidic a solution
becomes, the higher the number of H^+ and the lower the number of OH^-. The more
basic the solution, the lower the number of H^+ and the higher the number of OH^-.

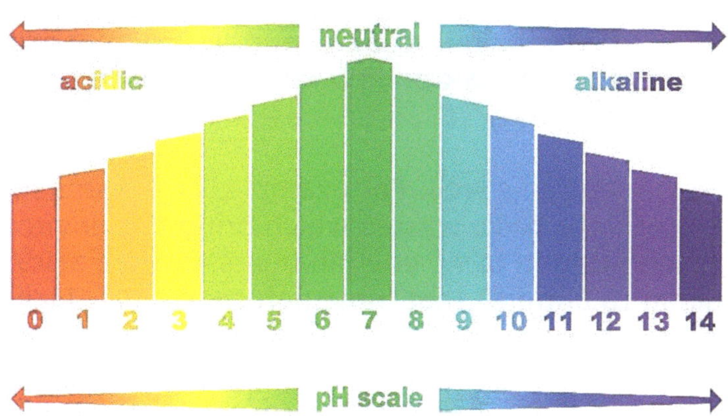

© Robin Atzeni/Shutterstock.com

An **acid** is a substance that causes an increase in the number of H^+ and a decrease
in the number of OH^- in a solution. The increase is usually the result of the acid
dissociating to release H^+. Some common acids and their products are:

Hydrochloric acid $\quad\quad HCl \quad \rightarrow \quad H^+ + Cl^-$
Acetic acid $\quad\quad\quad\quad CH_3COOH \quad \rightarrow \quad H^+ + CH_3COO^-$

A substance itself does not have to give up H^+ to cause an increase in the H^+ in a solution. For example, CO_2 combines with H_2O to form carbonic acid (H_2CO_3), which dissociates to form bicarbonate ion (HCO_3^-) and H^+.

$$CO_2 + H_2O \rightarrow H_2CO_3 \rightarrow H^+ + HCO_3^-$$

Therefore, CO_2 can result in an acidic solution when it is mixed with H_2O.

A **base** is a substance that causes a decrease in the number of H^+ in a solution and an increase in the OH^-. Some common bases that dissociate to produce OH^- are:

Sodium hydroxide $\qquad$ NaOH $\quad \rightarrow Na^+ + OH^-$
Magnesium hydroxide $\qquad$ $Mg(OH)_2 \quad \rightarrow Mg^+ + 2\,OH^-$

Ammonia (NH_3) dissolved in water is also basic. It does not directly produce OH^- but it can remove H^+ from a solution.

$$NH_3 + H_2O \rightarrow NH_4^+ + OH^-$$

The bicarbonate ion (HCO_3^-) is also basic. It can accept H^+:

$$H^+ + HCO_3^- \rightarrow H_2CO_3$$

Distilled water is neither an acid nor a base and has a pH of 7 which is neutral. We can also produce a neutral solution when we mix an acid and a base as long as the acid and base have neutralized each other. All of the H^+ donated by the acid would be absorbed by the base.

A **neutralization reaction** occurs when an acid combines with a base to form a **salt and water**. An example is the following:

$$HCl + NaOH \longrightarrow H_2O + NaCl$$

If you spill an acid or base in lab and want to clean it up safely, neutralizing an acid with a base (or vice versa) can help in the cleanup, but strong acids and bases can react explosively. **Always use weak or low concentration acids or bases for neutralization reactions.**

Neutralization occurs in nature as well. For example, organisms living in very acidic environments tend to excrete basic wastes which bring the environment back into equilibrium, while those living in basic environments excrete acids. This is one of the reasons (in addition to buffers) that most natural waters, including septic tanks and wastewater treatment ponds which have been allowed to work for some time, tend to have a pH near 7.

Anthocyanins are plant pigments responsible for red, blue and purple colors in flowers, fruits and autumn leaves. **Indicators** are chemicals that change color depending on the pH of the solution. Boiling red cabbage extracts these pigments which then can be used as a pH indicator.

To use an indicator to determine the pH of solutions, you must first determine the color changes that take place when the indicator is used with substances with known pH values. The color changes that occur in the cabbage extract when it is mixed with substances of various pH values will be used as a set of standards. You can then determine the pH of various substances by mixing each with the cabbage extract and comparing the resulting colors with the colors of the standards.

© Viktar Malyshchyts/Shutterstock.com

Procedure 2 Making the pH Standards

1. Use a marker to label 3 test tubes 4, 7, 10. Mark each of the tubes at the 1 cm and 2 cm line.

2. Place the appropriate buffer into each of the 3 tubes to the 1 cm line. (Place the pH 4 buffer into the tube 4 and so on).

3. Add the cabbage extract *very carefully* to each of the 3 tubes, up to the second line (the 2 cm line). Cap the tubes. **DON'T TOUCH THE PIPETTE THAT CONTAINS THE CABBAGE JUICE TO THE SOLUTIONS IN THE 3 TUBES**.

4. Vortex the tubes using the vortex mixer.

5. Record the color of each tube below and <u>save these tubes for reference in procedure 3</u>. Do not discard.

pH	Color
4	
7	
10	

Source: Werner Williams

Procedure 3   Measuring pH with the Cabbage Indicator

1. Label 2 clean test tubes A and B and mark them at the 1 cm and 2 cm lines.

2. Add solution A to the first mark of tube A and solution B to the first mark of tube B.

3. Add cabbage extract to the second mark. Cap the tubes and vortex to mix.

4. Compare the color in tubes A and B with the colors of the standards (from procedure 2).

pH of tube A_____________________; pH of tube B_______________________

Procedure 4 Measuring pH with Alkacid Test Paper

The cabbage indicator method is most accurately used with colorless or white solutions. **Alkacid test papers**, which contain indicators, are another means of measuring pH and can be used with colored solutions.

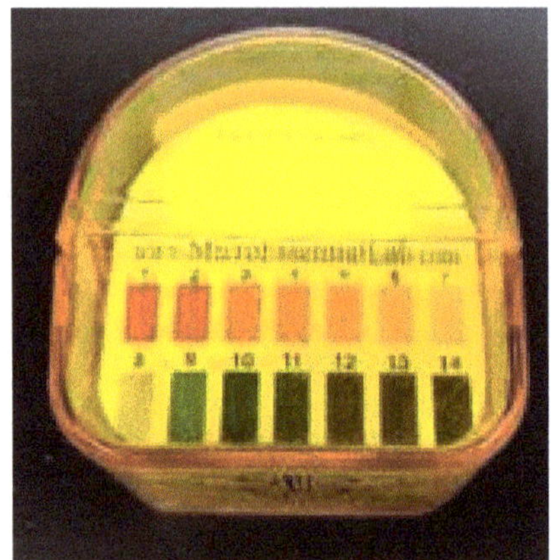

Source: Werner Williams

1. Tear off a small piece of test paper. Hold the test paper with a forceps. Dip a piece of the paper into solution A.

2. While the paper is still wet, compare its color with the standard pH color scale on the label of the alkacid paper's container.

pH of A ______________________

3. Repeat steps #1 and 2 with a <u>new piece of test paper</u>, using <u>solution B</u>.

pH of B ______________________

Do the results obtained with the alkacid papers match the results you obtained with the cabbage indicator for solutions A and B?

Explain any discrepancies.

<u>PROCEDURE 5 Determining the pH of Some Common Solutions</u>

1. Mark 3 tubes at the 1 cm and 2 cm marks.
2. Fill to the first mark with one of the solutions below and to the second mark with cabbage extract. Cap the tubes and vortex to mix.
3. Place your results in the table below.

Solution	pH
Sprite or 7-UP	
Ammonia	
Distilled water	

Source: Werner Williams

1. Which substance could be used as a control in procedure 5?

Enzymes function best at particular pH values. In the normal human stomach, a pH of 2.0 or 3.0 provides the environment required for those enzymes to work best. Milk of Magnesia, Maalox and Alka-Seltzer are often used to treat "acid-indigestion" of the stomach. This occurs when the pH of the stomach is even more acidic than 2.0, which interferes with the normal functioning of the enzymes.

2. Why should stomach medicines have pH values that are much higher than normal stomach pH which is around 2?

3. What might happen to the functioning of the enzymes if an excess of these medicines were used? Why?

<u>Using the pH Meter</u>
A **pH meter** measures the concentration of hydrogen ions (H+) in a solution. The meter's glass electrode contains an HCl solution of a known concentration. The pH meter measures the difference in the H^+ concentration inside the electrode with that found outside the electrode and converts that difference into a pH reading.

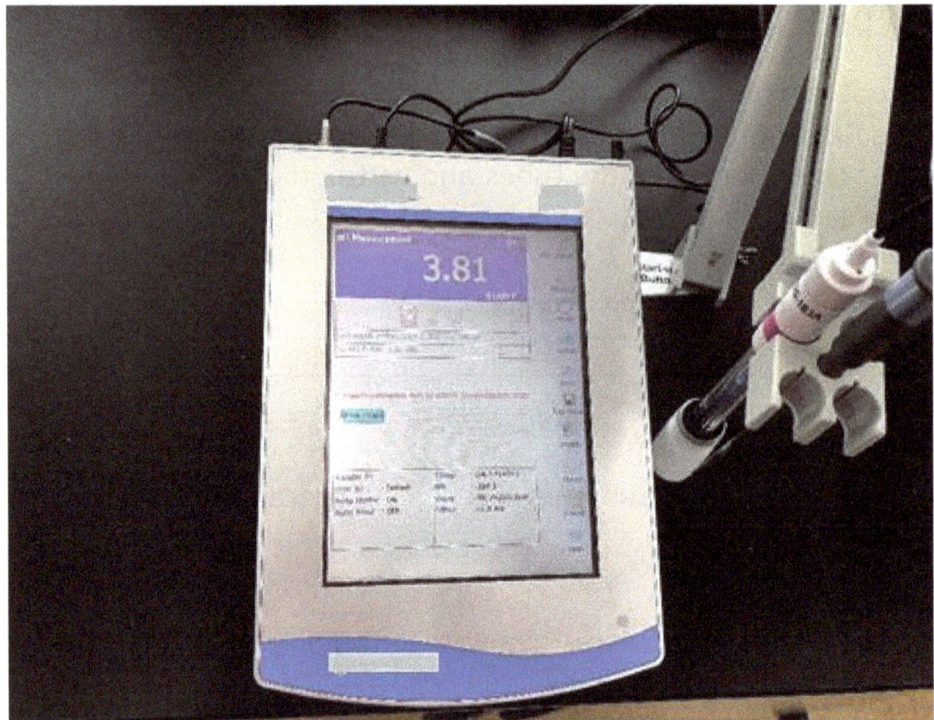

Source: Werner Williams

In this procedure, you are to determine the pH of solution A and solution B.

<u>Procedure 6 Using the pH meter</u>

1. Raise the pH electrode out of its storage beaker. **<u>(Be careful with the fragile glass electrode)</u>**. Over the "waste container," **rinse the electrode with distilled water from the wash bottle.**
2. Lower the electrode into the container of Solution A.
3. Use the pH stylus to **<u>gently</u>** tap the <u>measure</u> icon on the right of the display screen.
4. The pH reading should appear on the top of the screen. Record the pH in the table below.
5. Repeat instructions #1-4 with solution B.
6. Rinse the electrode with distilled water over the waste container and place it in the storage container.

Solution	pH
Solution A	
Solution B	

Source: Werner Williams

1. Compare the cabbage juice derived pH values with the derived pH values
   from the pH meter? Which is more accurate and why?

2. Is the H^+ conc. of a pH 3.8 solution higher or lower than that of a solution
   with a pH of 6.8?

Determining the Buffering Capacity

In order for biological systems to function normally, the pH of those systems
usually needs to be maintained in a narrow range. If there is an excess or deficit of
H^+ or OH^- outside of that range, it may interfere with the reactions that would
normally occur. For example, the pH of blood is usually maintained at about 7.4. If
it rises above 7.8 or drops below 6.8, the results could be deadly. Acidosis is a
condition in which the pH drops too low and alkalosis is a condition in which the pH
rises too high. Many of the reactions that take place routinely in the body can lower
the blood pH. Exercising and eating certain foods may have the same effect. To
counteract this lowering (or the raising) of the pH, the body has buffering systems
to help keep the pH constant.

A buffer is a substance that is made up of a weak acid and a weak base. Its pH
changes very little when a small amount of a strong acid or base is added to it. It
takes H^+ out of a solution (when it acts as a base) and releases H^+ into a solution
(when it acts as an acid). Its purpose is to maintain a constant pH. For example, a
pH 6 buffer maintains the pH at 6 and a pH 13 buffer maintains the pH at 13. The
goal of a buffer is not to make the pH neutral but to prevent sudden pH changes.

The buffering capacity of a solution is tested by adding small amounts of an acid or
a base, while monitoring the pH. If the pH changes dramatically, that substance is
not a good buffer. If the pH changes only slightly, it is considered to be a good
buffer. However, if we keep adding an acid or a base to the buffer, eventually its
ability to be a buffer ends, and its pH at that point will change drastically.

Look at Figure 1 below. Is substance A a good buffer? Explain.

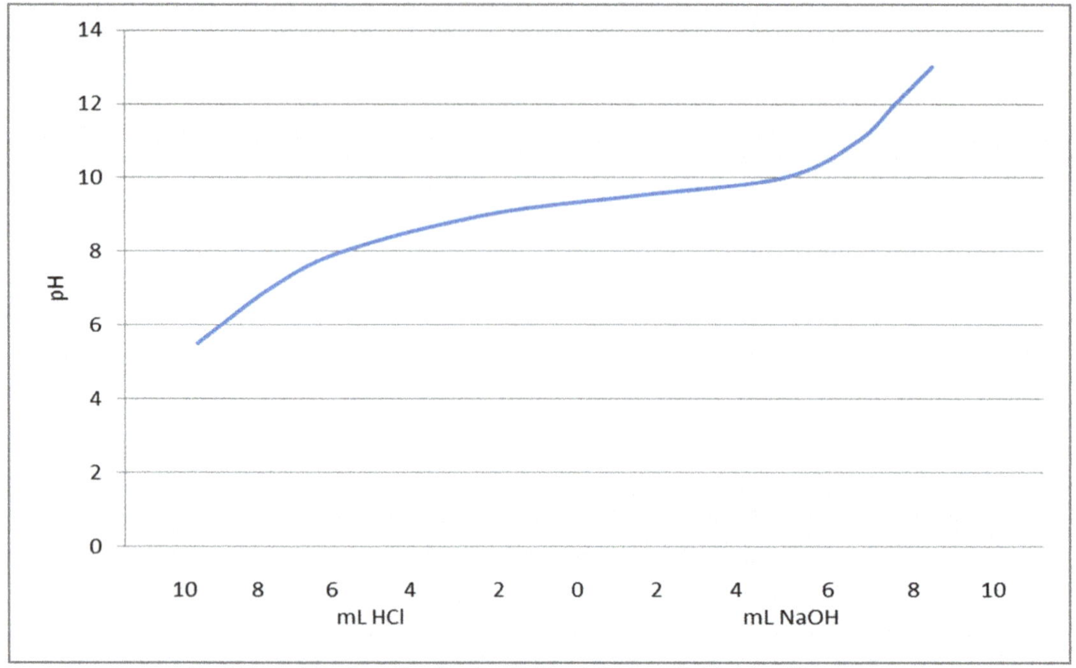

Contributed by Jennifer Raabe © Kendall Hunt Publishing Company

Figure 1 Titration curve for substance A

At what pH does substance A buffer?

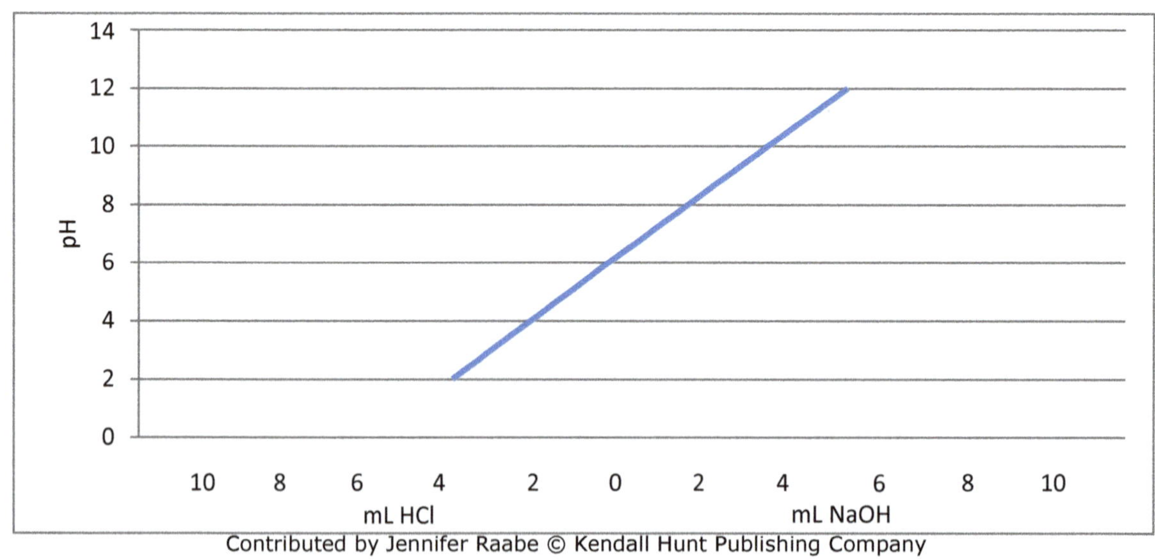

Contributed by Jennifer Raabe © Kendall Hunt Publishing Company

Figure 2 Titration curve for substance B

Is substance B a good buffer? Explain.

<u>Procedure 7 Determining the Buffering Capacity</u>

1. Your instructor will provide you with 2 flasks that contain a specific buffer solution to measure. Choose from the following buffers: pH 5 or 8.
2. Determine the pH of the solution by raising the electrode out of its storage solution and rinsing the electrode with the wash bottle of distilled water over the beaker that is labeled "Waste Water."
3. Swirl the flask with the buffer solution and **<u>carefully </u>**lower the electrode into the solution.
4. Take the pH reading and record it at 0 on the x-axis of Figure 3.
5. Use a micropipette to add 2.0 ml of HCl to your flask and swirl (about 10 times) to mix.
6. Record the new pH as "2 ml HCl added" on the X-axis in Figure 3.
7. Add another 2 ml of HCl and record the new pH as "4 ml HCl added" on the x- axis in Figure 3.
8. Continue adding 2 ml of HCl at a time and record the pH until you have added a total of 10ml or there is a significant decrease in pH, whichever comes first.
9. Raise the electrode out of the solution, rinse **<u>thoroughly </u>**with distilled water and wipe it with a Kimwipe.
10. Discard the used micropipette tip and replace it with a new tip.
11. Repeat instruction steps #3-8, using the second flask, but this time add 10ml of NaOH in 2 ml increments. Record the pH readings beginning with "2ml NaOH added" in Figure 3.
12. Raise the electrode out of the solution, rinse **<u>thoroughly </u>**with distilled water and wipe it with a Kimwipe.
13. Obtain 2 distilled water flasks.
14. Discard the used micropipette tip and replace it with a new tip.
15. Repeat instruction steps #3-8, using one flask of distilled water to determine its buffering capacity. Add HCl to the flask in 2 ml increments and record the results in Figure 4.
16. Raise the electrode out of the solution, rinse **thoroughly** with distilled water and wipe it with a Kimwipe.
17. Discard the used tip and replace with a new tip.
18. Repeat instructions #3-8 with the second flask of water. Add NaOH to the flask in 2 ml increments and record the results in Figure 4.
19. When finished, rinse the electrode **thoroughly** with the wash bottle of distilled water.
20. Place the electrode in its storage solution.

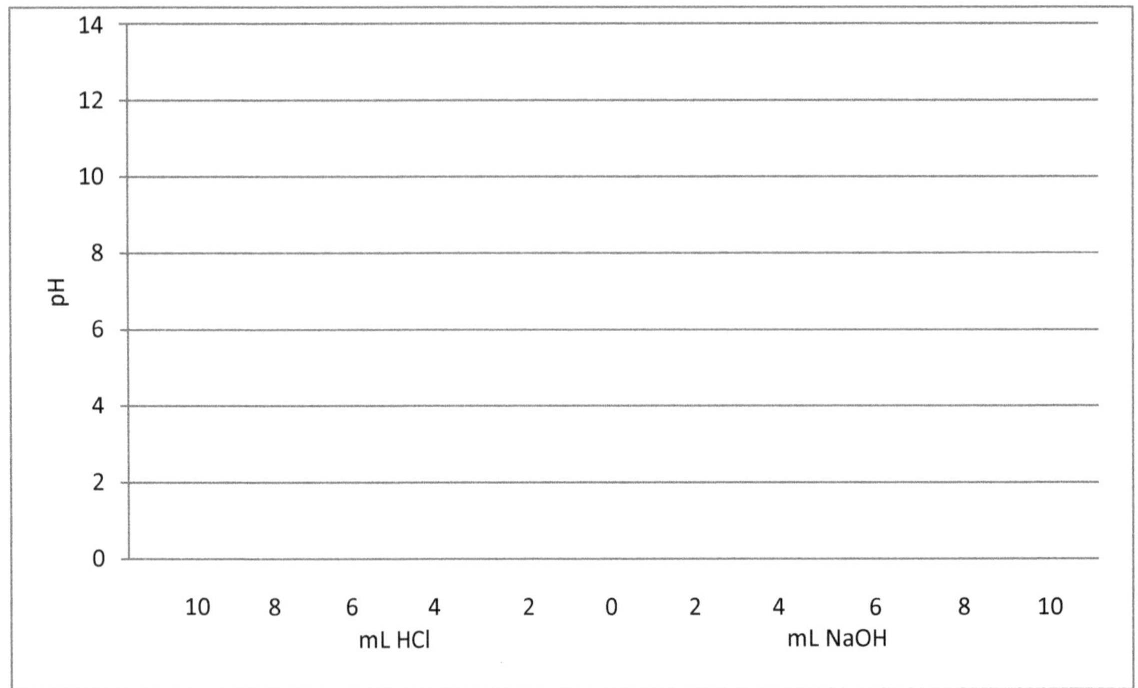

Contributed by Jennifer Raabe © Kendall Hunt Publishing Company

Figure 3 Titration Curve for pH__________Buffer

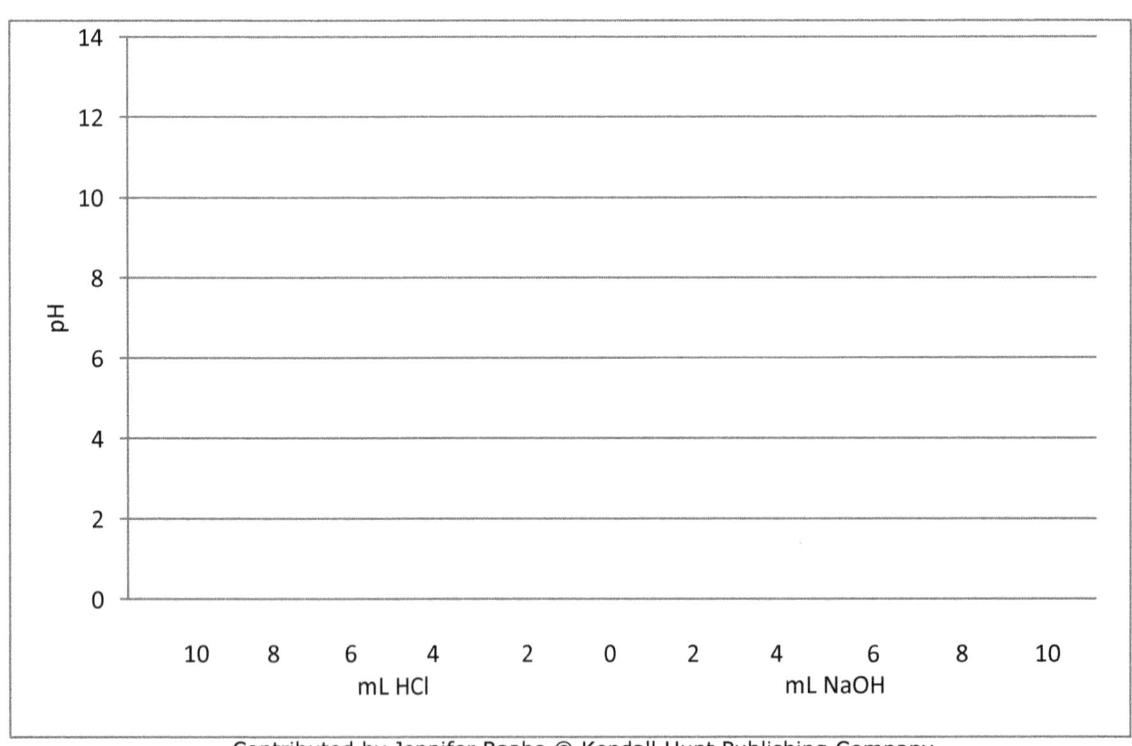

Contributed by Jennifer Raabe © Kendall Hunt Publishing Company

Figure 4 Titration Curve for Distilled Water

Questions
1) At what pH does your solution buffer?

2) Describe the buffering capacity (or lack of it) of your solution?

3) Is water an effective buffer?

4) Give an example of two solutions that when mixed together will produce a
   neutral solution?

5) When an acid and a base are combined in a neutralization reaction, what are
   the products of the reaction?

6) If you measure the pH of vinegar and find that it is 2 and black coffee is 5.
   How much greater is the H^+ conc. in vinegar than in the coffee?

7) How can you tell if a substance is a good buffer?

Lab 5: Detecting Organic Molecules

All living organisms are composed of various types of organic molecules such as carbohydrates, protein, lipids, nucleic acids, and vitamins. Various biochemical indicator tests may be used to determine the presence of such substances in a food sample. Indicator chemicals can show if these organic compounds are present in a food sample by a change in color. In this lab exercise you will perform several of these tests on both known and unknown substances. For each test performed, there will be a negative control (a test which gives a negative color result) and a positive control (a test which gives a positive color result).

A. Benedict's Test for Simple Sugars

Benedict's reagent is used as a simple test for sugars. The test is primarily for <u>monosaccharides</u> or single ring sugars. <u>Disaccharides</u> are double ring sugars and <u>polysaccharides</u> are many rings strung like beads on a string.

A single ring sugar such as glucose will open to the linear form, many showing a free aldehyde group. When heated in Benedict's reagent, which contains copper, the aldehyde group donates electrons to the Copper ions to form a green to red color, depending on the amount of sugar present.

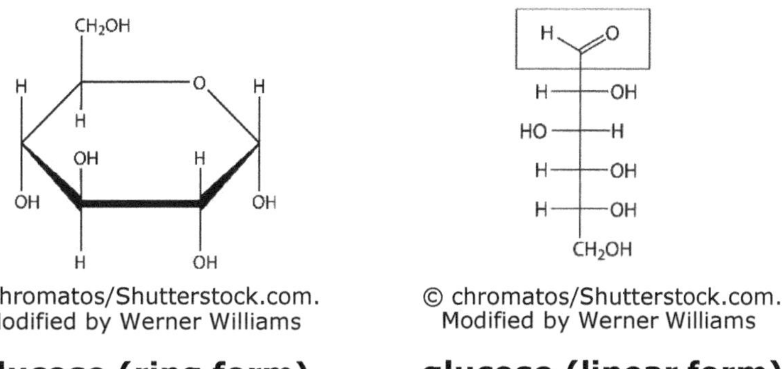

© chromatos/Shutterstock.com.
Modified by Werner Williams

© chromatos/Shutterstock.com.
Modified by Werner Williams

glucose (ring form)   **glucose (linear form)**

Benedict's reagent will be used to determine the presence of a sugar in various substances and to determine the relative amount of the sugar present. Record your results in the data table provided. Note that "results" mean what you actually <u>see</u>. The "conclusion" is what you deduce based on your results. The mixture will turn green if a <u>trace</u> of the monosaccharide sugar is present and yellow if a <u>small</u> amount of sugar is present. An orange color results if a <u>moderate</u> amount of a sugar is present and red if a <u>large</u> amount is present. For example, if a tube with a food substance and Benedict's reagent turns orange upon heating, you would indicate in results "orange." Your conclusion would be "moderate." If the color in the tube does not change (remains blue), it indicates a negative test.

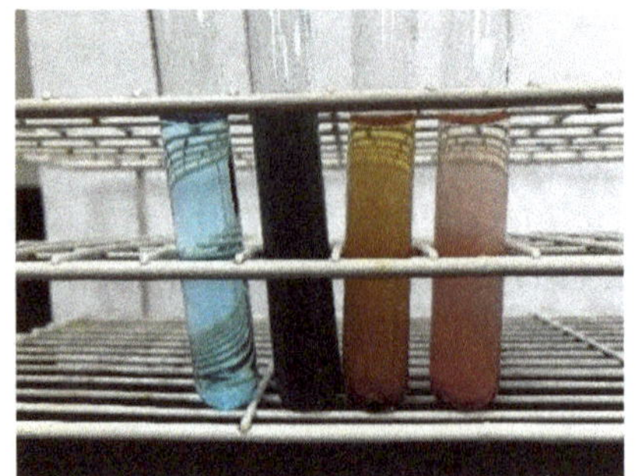

Source: Werner Williams

Blue-"Negative"; Green-"Trace"; Yellow-"Small"; Orange-"Moderate"; Red-""Large"

Materials	Test tubes	Marker
	Test tube racks	Metric ruler
	Test tube holder	Substances for testing
	Hot water bath	Benedict's reagent

<u>Procedure 1</u>

1. Use five test tubes from the rack and number them (1-5) with a marker. Using the metric ruler, place a line on each test tube **1 cm from the bottom** of the tube.
2. Fill each test tube up to the line with the various substances listed in the table below.
3. Add one **dropper full** (not one drop) of Benedict's solution to each tube. Shake gently to mix the solutions.
4. Using a test tube holder, carefully place all tubes in the hot water bath and allow to heat for approximately 1-2 minutes. Do not overheat.
5. After heating, be sure to use the test tube holder to remove the tubes from the hot water bath. **Caution: tubes will be hot.**
6. Observe the colors of the tubes and complete the table below.

Test tube	**Substance**	**Results (colors)**	**Conclusion (amount)**
#1	Water		
#2	5% glucose		
#3	5% fructose		
#4	5% sucrose		
#5	Starch solution		

Source: Werner Williams

Which tube serves as a negative control for simple sugars? _______________________

Which tube serves as a positive control for a simple sugar? _______________________

How does the structure of the sugar sucrose differ from that of glucose or fructose?

____

B. Iodine Test for Starch

<u>Iodine is used to test for</u> the presence of starch. Long chains of glucose molecules may be arranged in various chains to make starch molecules. Starch is a storage form of glucose in organisms such as plants. Starch can separate into two fractions: <u>amylose</u>, the linear form (10-20%), and <u>amylopectin</u>, the branched form (89-90%). In the presence of starch, iodine turns from brownish-red to deep blue or blue-black. Amylose is responsible for this color change as the iodine molecule slips inside of the amylose coil. Animals produce a similar glucose storage compound called glycogen. Since <u>glycogen</u> is a branching molecule, it does not show the color change with iodine.

Contributed by Kim Hakala. © Kendall Hunt Publishing Company

Amylose   Amylopectin

Contributed by Kim Hakala. © Kendall Hunt Publishing Company

Glycogen   Iodine complexes within amylose

Materials   Spot plate
Iodine solution
Substances for testing

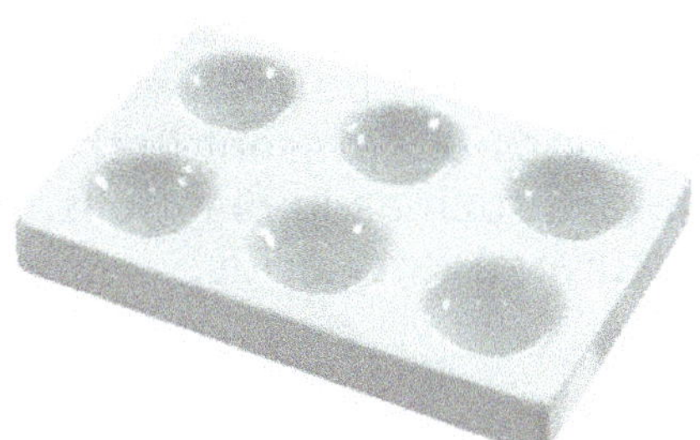

© Peter Sobolev/Shutterstock.com   Source: Werner Williams

Spot Plates

<u>Procedure 2</u>

1. Place a small amount of each of the substances listed below on a spot plate.
2. Add 1-2 drops of iodine solution to each substance on the plate.
3. Note any color changes and record your data and conclusions in the table below.

Substance	Results (colors)	Conclusion (starch present or not present)
Water		
Cornstarch		
Glucose		
Flour		
Potato		
Vegetable oil		

Source: Werner Williams

Which substance serves as a negative control for starch? ________________________

Which substance serves as a positive control for starch? ________________________

C. Sudan IV Test for Lipids

Lipids are non-polar molecules that are insoluble in water. Fats and oils are types of lipids in which a glycerol molecule is attached to one, two, or three fatty acids. These are called monoglycerides, diglycerides, or triglycerides respectively.

If the fatty acid contains only single bonds, the lipid will often be solid at room temperature such as in most animal fats. If double bonds are present in the fatty acids, the lipid will often be liquid at room temperature such as in plant oils. Some lipids may contain a mixture of saturated and unsaturated fatty acids.

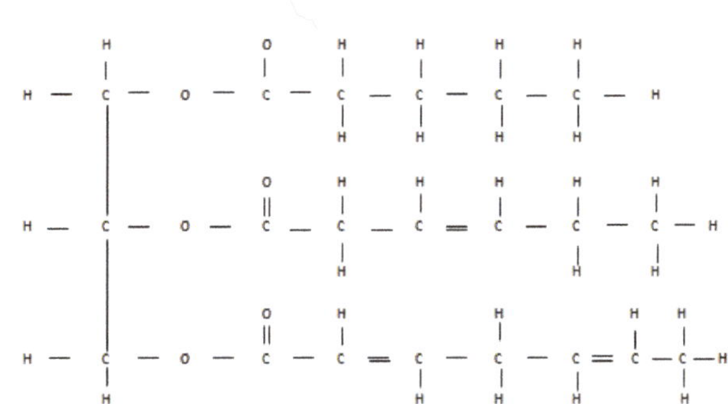

The diagram above shows a triglyceride with two unsaturated fatty acid and one saturated fatty acids.

Sudan IV stain is red colored. It is soluble (absorbed) in non-polar substances such as lipids and is useful as an indicator stain to detect lipids. Since a lipid is less dense than water and insoluble in water, the lipid will rise to the top of water in a test tube. Following the procedure, a dark pink/red color on the top of the test tube indicates a positive test for lipids and the water beneath the lipid remains clear. A pale pink or no color result in the test tube would indicate a negative test for lipids.

Materials: Test tubes   Marker
                          Test tube rack   Metric ruler
                          Test tube holder   Substances for testing
                          Sudan IV stain

Source: Werner Williams

Procedure 3

1. Use six test tubes from the rack and number them (1-6) with a marker. Using a metric ruler, place a line on each test tube **2 cm from the bottom** of the tube and another line on each test tube **4 cm from the bottom**.
2. Fill each test tube up to the 2 cm line with various substances listed in the table on the next page.

3. Add water up to the 4 cm line in each test tube.
4. Add three drops of Sudan IV stain to each tube. Shake gently to mix the solutions.
5. Let the tubes sit for three minutes.
6. Observe the colors and record your results (color observations) and conclusions in the table below.
7. For substances which test positive, indicate whether the lipid is more likely to contain single bonds or double bonds.

Substance	Results (colors)	Conclusion (lipid present or not present)	double bonds or single bonds
Water			
Olive oil			
Butter			
Corn starch solution			
Canola oil			
Glucose solution			

Source: Werner Williams

Which of the substances serves as a negative control for lipids? ___________________

Which of the substances serves as a positive control for lipids? ___________________

D. Biuret's Test for Protein

Amino acids are joined together to form proteins by means of peptide bonds. Any compound with more than two peptide bonds will react with biuret reagent, which contains copper sulfate ($CuSO_4$) mixed with sodium hydroxide (NaOH). The amino group (NH_2) of amino acids forms a complex with the copper ions of the biuret's reagent and causes the change in color. In a positive Biuret's test, the mixture will turn from blue to purple. As two peptide bonds are required for the formation of the complex, single amino acids and dipeptides with only one peptide bond will give a negative test.

Source: Werner Williams

The diagram below illustrates a tri-peptide with three amino acids and two peptide bonds.

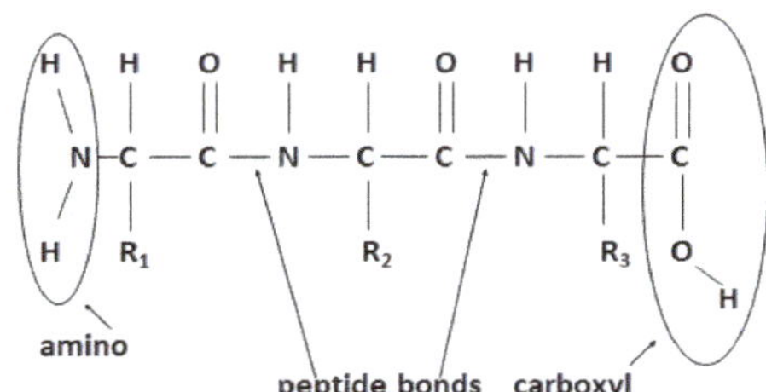

Contributed by Kim Hakala. © Kendall Hunt Publishing Company

The diagram below illustrates the complex between copper ions (Cu^{2+}) of the biuret reagent with peptide bonds of a protein.

Contributed by Kim Hakala. © Kendall Hunt Publishing Company

Materials Spot plate
Biuret's solution
Toothpicks
Substances for testing

<u>Procedure 4</u>

1. Place a small amount of each of the substances listed below on a spot plate.

2. Add 1-2 drops of Biuret's solution to each substance on the plate. You may need to stir mixtures with a toothpick.

3. Note any color changes and record your data and conclusions in the table below.

Substance	Results (colors)	Conclusion (protein present or not present)
Water		
Albumin (powdered egg white)		
Glucose		
Cornstarch		
Oil		
Plain yogurt		

Source: Werner Williams

Which of the substances serves as a negative control for proteins? _________________

Which of the substances serves as a positive control for proteins? _________________

E. Test for Vitamin C

Vitamins are another class of organic molecules found in living cells. Unlike the other classes, they do not have building block subunits. The indicator indophenol is dark purple. Vitamin C is a strong reducing agent and is easily oxidized. In the presence of vitamin C, indophenol solution will be reduced and will completely lose its color.

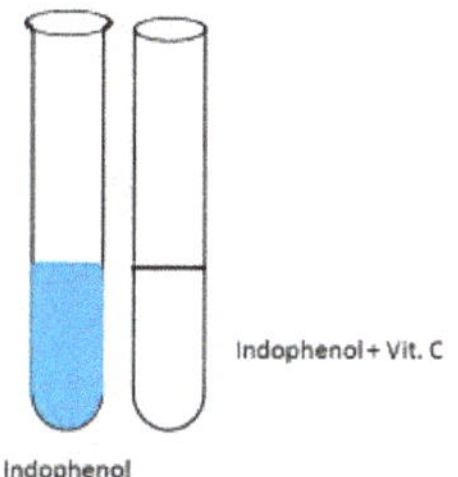

Source: Werner Williams

Contributed by Kim Hakala. © Kendall Hunt Publishing Company

Oxidation of Vitamin C

Materials   Spot plate
Indophenol solution
Toothpicks
Substances for testing

Procedure 5

1. Place a small amount of each of the substances listed below on a spot plate.

2. Add 4 drops of indophenol solution to each substance on the plate. You may need to stir mixtures with a toothpick.

3. Note any color changes and record your data and conclusions on the table below.

Substance	Results (color remains or color disappears)	Conclusion (vitamin C present or not present)
Water		
Vitamin C solution		
Glucose		
Lemon Juice		

Source: Werner Williams

Which of the substances serves as a negative control for vitamin C? _________________

Which of the substances serves as a positive control for vitamin C? _________________

F. Unknowns/ Foods:

Your instructor will supply you with various unknowns or foods to be tested.
Perform each of the organic molecule tests to the two foods that are assigned to
your group. Record your results and conclusions in the data table below. Your group
will be assigned two different foods to test but you are responsible for the results of
foods tested by all lab groups.

Foods	Simple Sugar		Starch		Lipid		Protein		Vitamin C	
	Results	Conclusion	Results	Conclusion	Results	Conclusion	Results	Conclusion	Results	Conclusion
1. Tuna										
2. Peanut Butter										
3. Milk										
4. Refried Beans										
5. Lettuce										
6.										

Source: Werner Williams

Lab 6: The Use and Care of the Microscope

Many biological objects are so small that a microscope is needed to see them. Light microscopes use light rays that are focused using magnifying lenses.

How a Microscope Works

Light is released from a light source and is directed by a condenser lens onto the specimen which is usually found on a slide. The light from the specimen then passes into an objective lens, which magnifies and bends the light rays and sends it to a projector lens, which reverses the direction of the rays so that when it reaches the eye, it will not appear to be upside down. Some microscopes don't have a projector lens, so the person may be seeing a reverse image. You will know you are seeing a reverse image because when the slide is moved, it will appear to be moving in the opposite direction to the viewer. As the light rays travel to the eyepiece the image is further magnified and projected into the eye. Figure 1

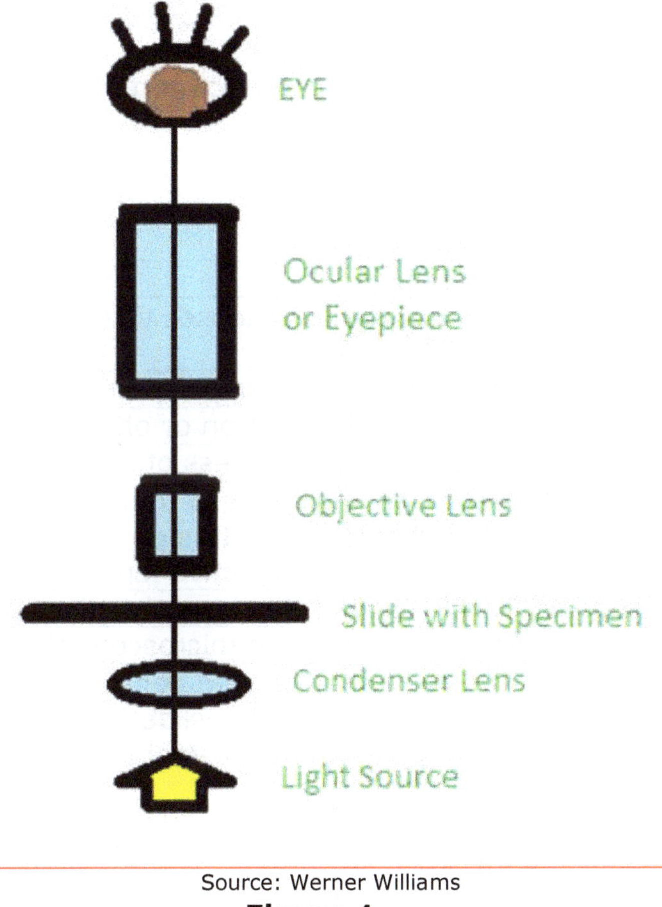

Source: Werner Williams
Figure 1

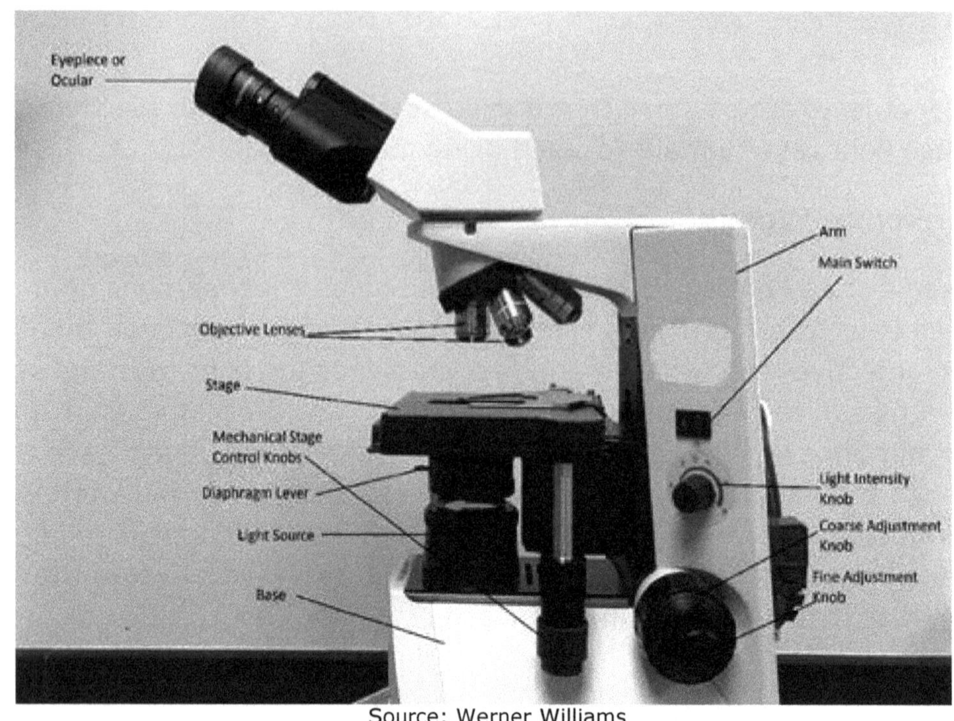

Source: Werner Williams

Figure 2 Parts of a Microscope

Table 1 Microscope Parts and Their Functions

Parts	Function
Eyepiece or Ocular	Topmost series of lenses through which we look to view the specimen.
Arm	This connects the upper parts to the base.
Main Switch	This turns the light on or off.
Light Intensity Knob	Controls the brightness of the light
Coarse Adjustment Knob	This large knob is used to bring the specimen into approximate focus. It is only used with the scanning and low power objective lenses.
Fine Adjustment Knob	This is used to fine tune the focus on the specimen.
Base	The flat part of the microscope that rests on the table.
Mechanical Stage Control Knobs	The two knobs located beneath the stage that control the movement of the slide. One controls the forward/reverse movement and the other controls the left/right movement.
Light Source	A lamp that sends a beam of light through the specimen
Condenser	A lens mounted usually below the stage that focuses or condenses the light onto the specimen
Diaphragm Lever	This lever varies the amount of light passing through the stage opening. This helps illuminate the specimen and helps in contrast and resolution.

Parts	Function
Stage	The flat platform on which the slide is placed.
Objective Lenses	The lenses closest to the specimen with the magnification values (4x, 10x, 43x, and 100x) engraved on the objectives. *Scanning*-Lens with the least magnification (usually 4x). *Low Power*- Lens with a greater magnification than the scanning objective (usually 10x). *High Power*- Lens with a greater magnification than the low power objective (usually 40x). *Oil Immersion*- Lens with the greatest magnification (usually 100x).
Revolving Nosepiece	The part of the microscope that holds the objective lenses.

Source: Werner Williams

Magnification

Compound microscopes contain at least two lenses. The first lens would be the **eyepiece**, also called the **ocular**, and usually magnifies objects 10 times (10X). The other lenses are the **objective lens**, which vary in their magnification. They usually range from 4X to 100X. The 100X lens usually called the oil immersion lens is used for viewing very small objects such as bacteria. Prior to using this lens, a drop of immersion oil is placed on the slide.

Contrast

Contrast refers to the darkness of the background relative to the specimen. There must be sufficient contrast between the different parts of the object you are viewing for you to distinguish those different parts. Sometimes stains are added to the specimen to increase the contrast. Reducing the amount of light improves the contrast when viewing unstained specimens.

Resolving Power

A good microscope not only magnifies objects but also allows us to **resolve** or distinguish objects. A blue light filter placed in the condenser or light source helps in distinguishing different objects. A good light microscope can resolve objects that are about 0.5μm or more apart

<u>**Focusing**</u>

As a general rule, you should start focusing with the low power lens (10X) and then use the high power objective lens (40X). Starting with a lower power lens allows you to more easily view and magnify a larger part of the specimen. When a larger power lens is used, you are actually viewing a smaller part of the specimen.
Figure 3

<u>**Working Distance**</u>

This is the space between the lens of the microscope and the top of the specimen. This is the free area available to work and manipulate the specimen. Of the three objective lens shown in Fig. 3, the 10X lens has the largest working distance and the 100X lens has the smallest.

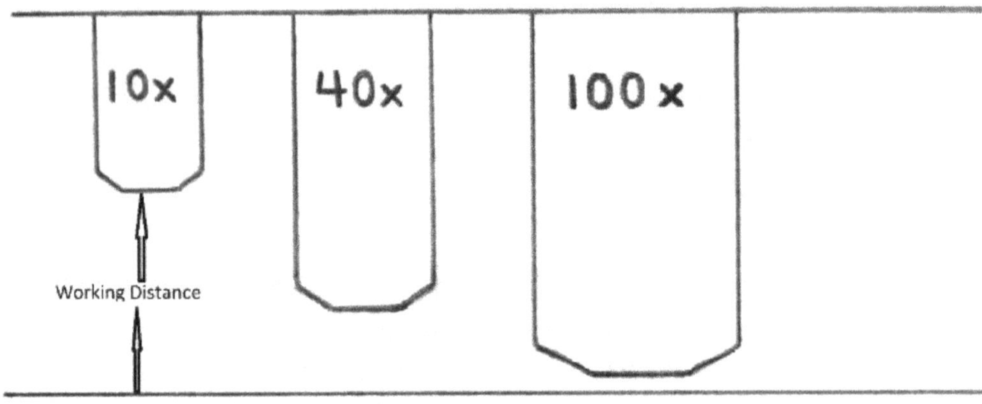

Source: Werner Williams

Figure 3

<u>**Parfocal**</u>

Your microscope is parfocal, meaning that if an image is in focus under low power and you wish to switch to a higher power, when you switch, it should still roughly be in focus. You should only have to make small adjustments with the fine adjustment knob to see a clear image.

<u>**Rules for Microscope Use**</u>

1. Use both hands when carrying the microscope.
2. The lowest power objective should be in position both at the beginning and end of microscope use.
3. Use only lens paper for cleaning the lenses.
4. Do not remove parts of the microscope.
5. Report any malfunctioning.

<u>**Total Magnification**</u>

The total magnification is calculated by multiplying the power of the ocular (5X or 10X) by the power of the objective lens.

Total Magnification = Ocular lens power x Objective lens power

Magnification of your ocular lens alone=

Complete the following table showing the total magnification using each objective.

Table 2 Total Magnification

	Magnification of Each Objective Lens Alone	Total Magnification (Objective x Ocular)
Scanning		
Low Power		
High Power		
Oil Immersion		

Source: Werner Williams

<u>**Diameter of the Field of View**</u>

The **field** is the lighted circle you see when you look through the eyepiece. You will use a ruler to help estimate the field of view.

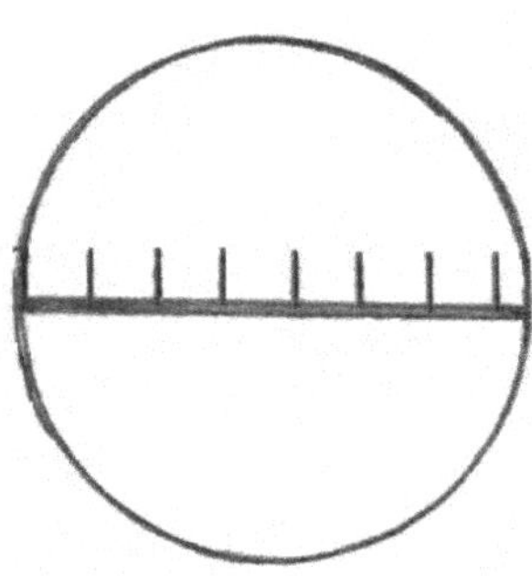

Source: Werner Williams

Figure 4 View of Ruler Through a Microscope

<u>Procedure 1 Diameter of the Field of View</u>

1. Place a clear plastic ruler on top of the metal clips that are used to hold a slide on the stage.

2. Place the millimeter scale over the center of the stage opening.
3. Click the 4X objective lens into place.
4. Use the <u>mechanical stage control knobs</u> to move the ruler so as to locate the millimeter lines of the ruler and place these lines in the middle of the field of view.
5. Use the coarse adjustment knob to bring the lines into focus. The distance between two lines on the ruler represents 1mm.
6. While looking through the eyepiece, set the mechanical stage control knobs to move the ruler so that one of the millimeter lines is just touching the left side of the field of view. Figure 4.
7. Count the number of millimeter lines (actually spaces) that are visible. Estimate to the nearest tenth of a millimeter. What is the diameter? Convert that into <u>micrometers (µm)</u>. Use the formula below:
   1mm=1000 µm;   So 1.3 mm=1,300 µm
   Scanning lens (4X, 40X total) field of view:___________mm or___________µm
8. Click the low power lens in place and repeat the instructions above (#4-7).
   Low-power objective (10X, 100X total) field of view:______mm or________µm

Because under high power (40x, 400x total) the thickness of one of the millimeter lines takes up practically the entire field of view, it is difficult to estimate the diameter of the field view under high power magnification.

The diameter under high power can be calculated on paper. Record the following data for your microscope:

1. LPD=low-power diameter of field (in micrometers)_____________
2. LPM=low-power total magnification (not scanning)._____________

3. HPM=high-power total magnification___________________
4. Compute the high-power diameter of field (HPD) by plugging in the data into the following formula:

$$HPD = LPD \times (LPM / HPM)$$

High-power objective (40x, 400 total field of view):_______________ µm

Does low power or high power have a larger field of view and allow you to see more of the object?_______________________________

Which has a smaller field but magnifies to a greater extent, the 4X, 10X, 40X or 100X?

Depth of Field

When we view objects with a microscope, we are viewing those objects from above. The distance between the closest and farthest objects in focus within a scene is called the depth of field. You will be using a slide with three colored threads similar to what you see in Figure 5.

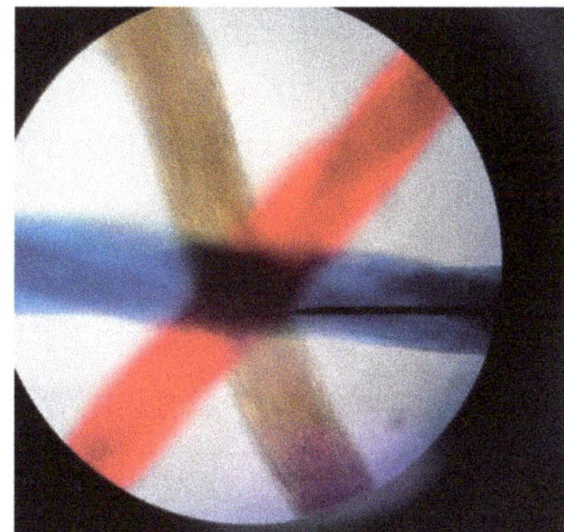

Source: Werner Williams

Figure 5 Three Colored Threads

Procedure 2 Depth of Field

1. Obtain a slide with three crossed colored threads and focus using the low power objective and the coarse adjustment knob.
2. Move the slide so that the area where all three threads overlap is in the center of your field of view and focus.
3. Switch to the high power lens and use the fine adjustment knob to fine-tune the image. *Never use the coarse adjustment knob with the high-power or oil immersion objectives.* Move the slide so you can see all three colors at once.
4. Looking through the microscope, turn the fine adjustment knob toward you, to lower the stage just until the threads are out of focus. Now, slowly roll the stage upward noting which color thread comes into clear focus first, then second, then last. The first thread to come into focus when you raise the stage is the one that is on top. The second thread to come into focus is the one in the middle.
5. Record your observations, relative to which color of thread is on the top, middle, and bottom.
   Top thread color _________________________________
   Middle thread color _________________________________
   Bottom thread color _________________________________

Remember, *never start observations with the high power or oil immersion objectives. Instead, start at a lower power and work up to the 40X or 100X objectives.*

<u>Procedure 3 Viewing the Letter "e"</u>

1. Obtain a microscope slide with the letter "e."
2. Place the slide on the stage with the letter over the circular opening in the stage. The slide should be placed so that "e" can be read with the naked eye.
3. Rotate the 4X objective into place and focus using the coarse adjustment knob.
4. Center the letter "e" and focus. Can you see all of the "e"? What is different about the orientation of the letter when viewed with the microscope instead of the naked eye? ____
____
5. Move the slide to the right while looking through the eyepiece. In which direction does the image move?_________Move the image to the left. In which direction does it move? ______________________________
6. Center the "e", focus and rotate the low power lens into place.
7. Center the "e" and focus. How much of the "e" can you see?________________
8. Rotate the 40X objective into place and focus using the fine adjustment knob. Remember *never use the coarse adjustment knob with the high power or oil immersion objectives*. You should be seeing "ink blotches" that compose the "e".
9. Draw the letter "e" when observed with each of the following:

    Unaided eye   4x objective   10x objective   40x objective

When you have been viewing a specimen under low power, and you switch to high power, does the image get larger or smaller? ______________________________________

When you have been viewing a specimen under low power, and you switch to high power, can you see more or less of the specimen? ____

Should you use low or high power when first looking at a slide under the microscope? Why?

____

Should you use the coarse or fine adjustment when looking at a slide under high power?

____

Preparing a Wet Mount

In a wet mount, a specimen is placed in a drop of water or other liquid and then covered with a coverslip. Figure 6. You will be performing two wet mounts; one of your cheek cells and the other of a slice of onion

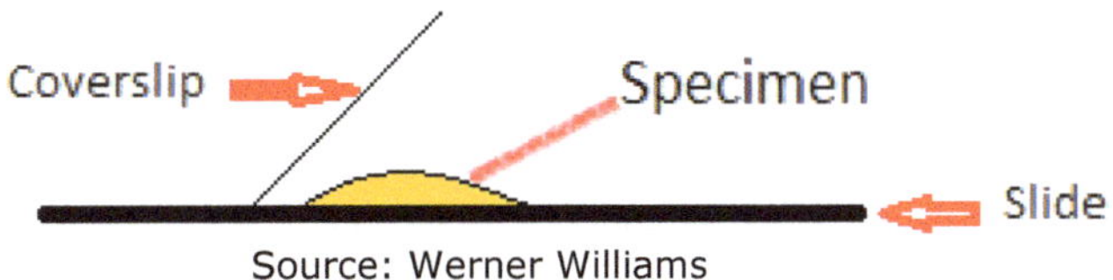

Source: Werner Williams

Figure 6 Wet Mount

Procedure 4 Wet Mount of Cheek Cells

1. Obtain a clean microscope slide and a cover slip.
2. Place a small drop of iodine on the slide.
3. Obtain a clean toothpick from your instructor and gently scrape the inside of your cheek about ten times.
4. Roll the toothpick with the cheek scrapings into the drop of stain on your slide and stir the contents gently.
5. Discard the tooth-pick.
6. Carefully lower a cover slip over the stain/cheek scrapings mixture. Note: slowly lowering the cover slip at about a 45°degree angle helps avoid air bubbles getting trapped beneath the cover slip.
7. Observe your cheek scrapings under the scanning lens (4X), the low power lens (10X), and the high power lens (40X). Remember to put the cells in the center of your field of view before increasing the magnification.
8. Find a few isolated cheek cells and focus on them. (Ignore clumps of cells). You should see roundish cells, with a stained nucleus in the center.
9. Draw a couple of your cells seen under high power below and then discard the slide and coverslip in the beaker of bleach.

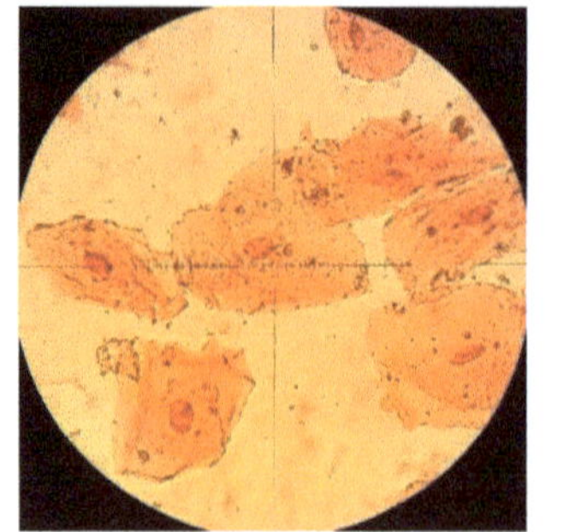

© Thawatchai Tumwapee/Shutterstock.com

Figure 7 Cheek Cells

<u>Procedure 5 Wet Mount of Onion Cells.</u>

1. Obtain a clean microscope slide and coverslip.
2. With a scalpel or your fingers, strip a thin, transparent layer of cells from a piece of onion.
3. Place it flat on the slide.
4. Add a drop of iodine and cover with a coverslip.
5. Observe the cells under the scanning lens (4X), the low power lens (10X), and the high power lens (40X). Be sure to label your drawing.
6. Locate the cell wall. Is a nucleus visible?_______ One to three nucleoli may be visible within the nucleus.

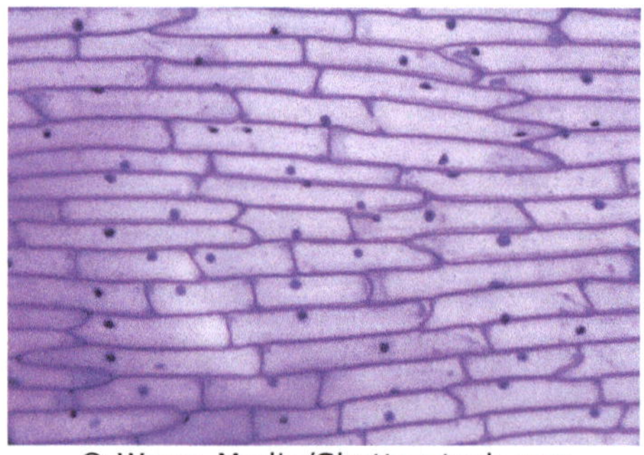

© Wave_Media/Shutterstock.com

Figure 8 Onion Cells

7. Draw a couple of the cells on high power and label visible structures.
8. Record some obvious differences between the human cheek cells and the onion cells. ____

____

9. Record any similarities between the two types of cells. ________________

____

<u>Procedure 6 Wet Mount of Aquatic Leaf Cells.</u>

1. Obtain a clean microscope slide and coverslip.
2. Pick a single leaf from the plant specimen provided.
3. Place it flat on the slide.
4. Add a drop of water from the plant container.
5. Carefully lower a cover slip over the leaf.
6. Observe the leaf under the scanning lens (4X), the low power lens (10X), and the high power lens (40X). Be sure to label your drawing.
7. Locate the cell wall. Locate the green colored chloroplasts. The nucleus may not be visible since a stain is not applied to the specimen.
8. Draw a couple of the cells on high power and label visible structures.

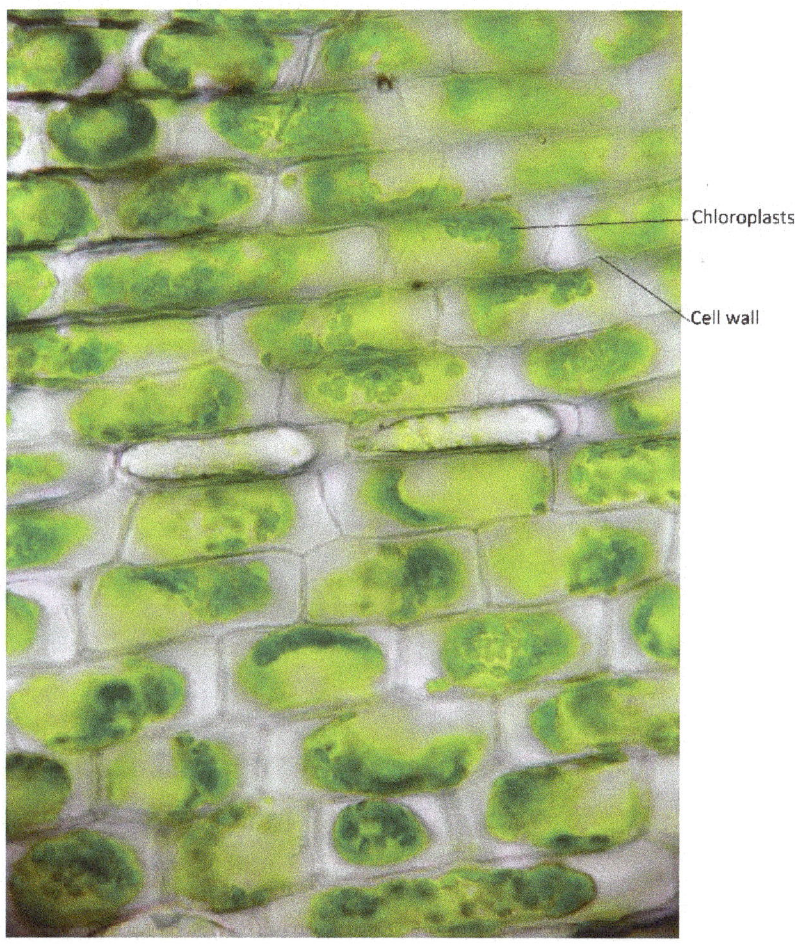

© Rattiya Thongdumhyu/Shutterstock.com. Modified by Werner Williams

<u>**Prokaryotic Cells (Demonstration Microscope)**</u>

Prokaryotic cells which include Bacteria and Archaea are simpler than eukaryotic cells. They do not possess a nucleus or other membrane-bound organelles and usually come in three basic shapes: round, rod-shaped and spiral.

A demonstration microscope has been set up with a slide of three different bacterial cells. In one field of view, you should be able to see at least two of the three different shaped cells.

Lab 7: Movement of Material Across Cell Membranes

The cell membrane of every cell is a **selectively permeable** barrier, controlling which molecules move into and out of the cell. For example, the cell takes in nutrients and oxygen and expels waste and carbon dioxide. The membrane also regulates the concentration of ions by transporting them one way or the other across the membrane.

Molecules are in constant motion, moving randomly (**Brownian movement**). **Diffusion** is one result of this motion, which is the net movement of particles from an area where their concentration is high to an area where their concentration is lower. This spontaneous process requires no input of energy.

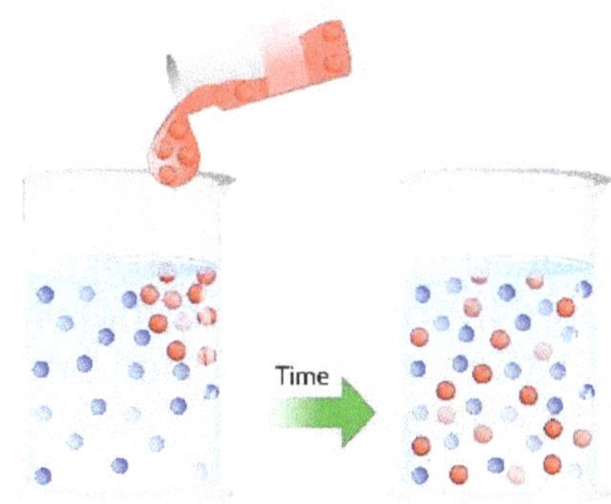

©DesignuaShutterstock.com

Diffusion

If a substance is more concentrated on one side of a membrane than on the other, it diffuses across the membrane from the region of higher concentration to the region of lower concentration, as long as the membrane is permeable to that substance. This is called **passive transport**, because it requires no energy to make it happen.

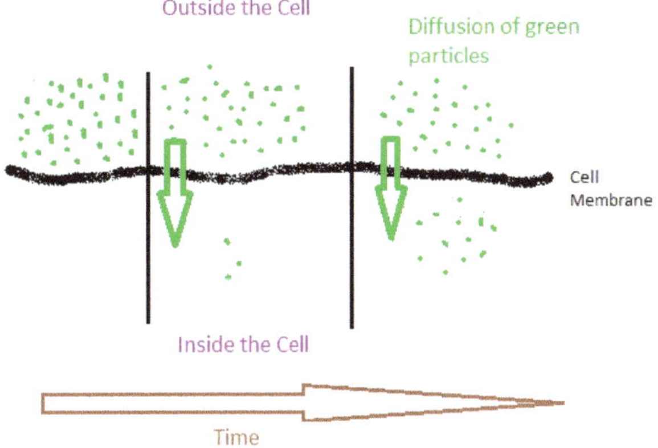

Source: Werner Williams

Water is one molecule that easily crosses the cell membrane. **Osmosis** is the net movement of *water* through a selectively permeable membrane from high concentration to low concentration. In a solution, the higher the concentration of solutes, the lower the concentration of free water molecules.

When a cell is placed in a solution in which the concentration of the solutes is lower than the solutes in the cell (and therefore, the concentration of water outside the cell is higher), then water will move into the cell. That solution is called **hypotonic**. As water molecules enter the cell, the cell will swell, especially if the solute cannot cross the cell membrane.

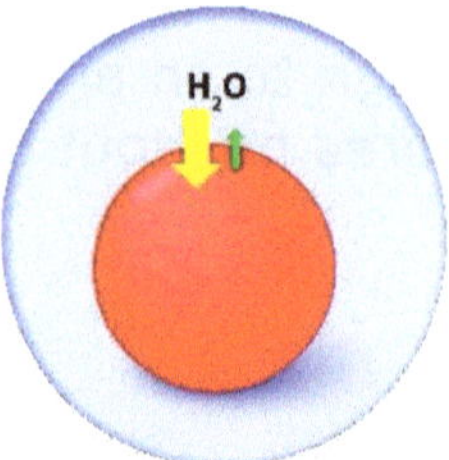

© Designua/Shutterstock.com. Modified by Werner Williams

If a cell on the other hand is placed in a solution in which the concentration of the solutes is higher than the solutes in the cell (and therefore, the concentration of water outside the cell is lower), then water will move out of the cell and the cell will shrink. That solution is called **hypertonic**.

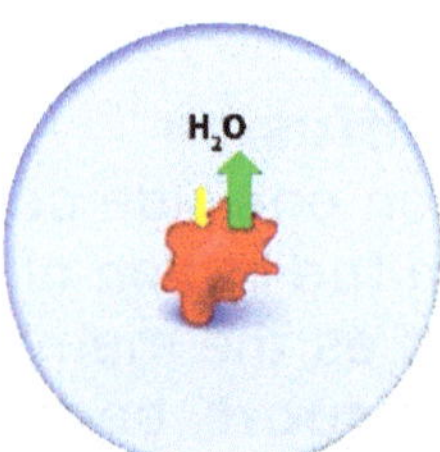

© Designua/Shutterstock.com. Modified by Werner Williams

A cell that is placed in an **isotonic** solution will have no net increase or decrease of water.

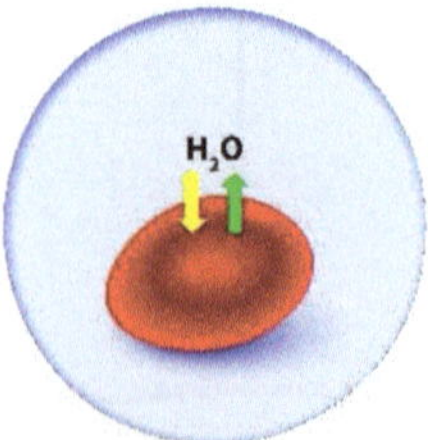

© Designua/Shutterstock.com. Modified by Werner Williams

Brownian Movement

In 1827, the English botanist Robert Brown was using his microscope to examine pollen grains that were suspended in water. He noticed that they jiggled about following a zigzag path. This zigzag motion of the particles is called **Browninan movement**, and is generated by the thermal motion of the molecules of the fluid. The higher the temperature of the liquid, the faster the movement of the particles.

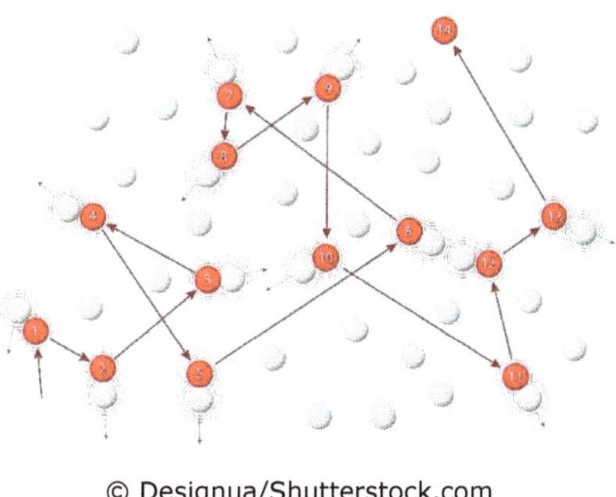

© Designua/Shutterstock.com

Figure 1. Brownian Movement

In today's lab, although individual molecules are too small to be seen, the random movement of the atoms and molecules in a fluid, colliding with carmine particles, can be observed.

Procedure 1-Browninan Movement

1. After **gently** shaking a bottle of carmine solution, place a drop on a slide, add a coverslip and observe the slide first under low power and then under high power.

2. Describe below how the carmine particles are moving in the water.

3. Place the slide in the beaker of bleach.

Osmosis Experiment

In this procedure, a dialysis tubing is used to represent a selectively permeable membrane. Since sucrose is unable to pass through the bag because of its size, you will be observing the movement of water either into or out of the bag.

<u>Procedure 2-Osmosis Into or Out of a Dialysis Tubing.</u>

1. Obtain a pre-soaked dialysis tubing (bag) and 2 pieces of string.
2. Fold the bag over onto itself at one end and tie a **secure knot** at that end using one string. Leave about ½ inch between the string and end of the bag.
3. Squeeze the other end of the bag between your fingers to open that end and using a transfer pipet, place the 30% sucrose into the open end until it is about 1/2 full.

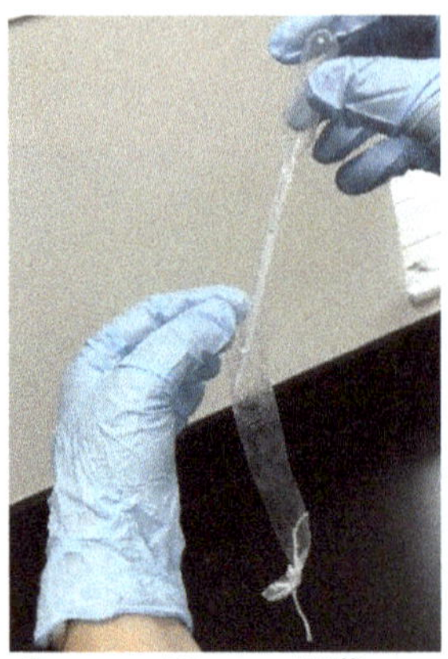
Source: Werner Williams

4. Squeeze the bag gently to remove excess air, fold the open end of the bag over onto itself and tie with string. The bag should not be so filled that it is tight or turgid. <u>Cut off the excess string.</u>
5. Rinse the outside of the bag under tap water and gently blot it dry with a paper towel.
6. Weigh your bag to the nearest 0.1g and place your bag in a beaker that contains either: 1) distilled water, 2) 10% sucrose or 3) 40% sucrose. (<u>Your instructor will assign you a specific concentration.</u>)
7. Place the bag in the beaker making sure that the bag is covered. Let it sit undisturbed for 30 minutes.

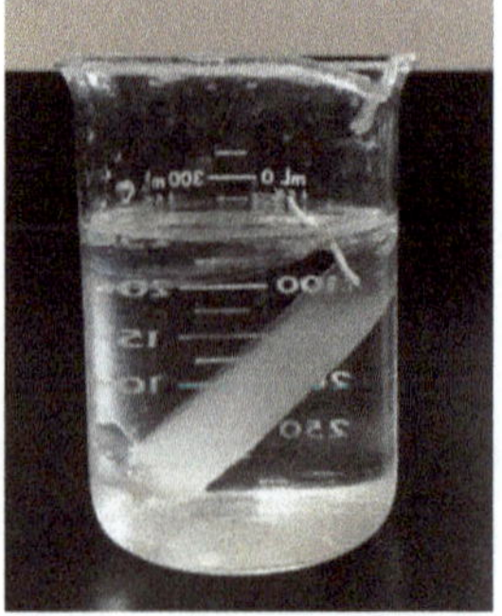
Source: Werner Williams

8. Remove the bag, <u>rinse, blot dry with a paper towel</u>, reweigh and record your data in Table 1.

Table 1. Weight of Dialysis Bags

	Distilled Water	10% Sucrose	40% Sucrose
Weight at time 0 min.			
Weight at 30 min.			
Change in weight (include + or -)			

Source: Werner Williams

1. Which solution is hypotonic to the cell, which is represented by the bag?

2. Which solution is hypertonic to the cell which is represented by the bag?

3. What would happen in the following three beakers if the bag was filled with distilled water instead of 30% sucrose?

<u>Distilled Water</u>   <u>30% Sucrose</u>   <u>60% Sucrose</u>

What would happen if the bag was filled with 60% sucrose?

<u>Distilled Water</u>   <u>80% Sucrose</u>   <u>60% Sucrose</u>

<u>Diffusion Experiment</u>

Starch is a storage form of glucose found in plants, glucose is a monosaccharide and iodine is an atom that contains 53 protons. In procedure 3, you will determine which of these three particles is large enough to move through the pores of a dialysis tubing.

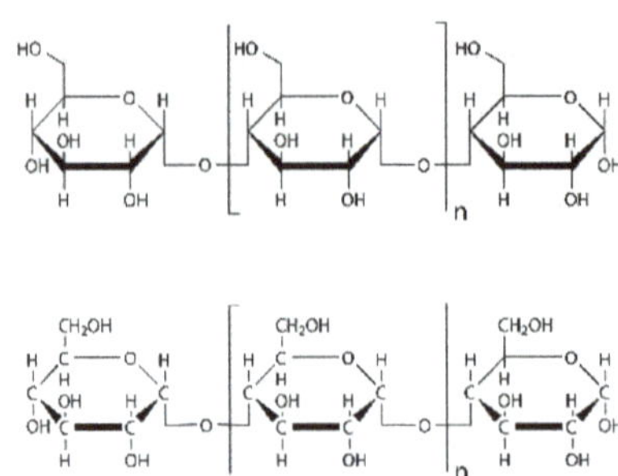

© chromatos/Shutterstock.com

Starch

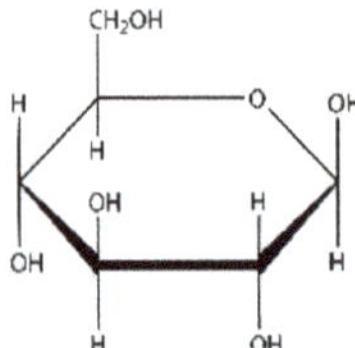

© chromatos/Shutterstock.com. Modified by Werner Williams

Glucose

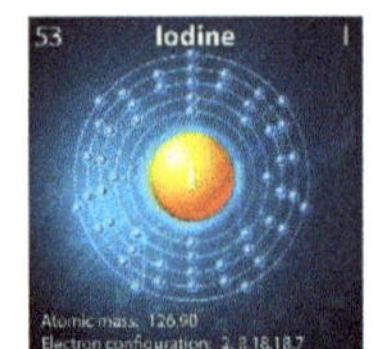

© BlueRingMedia/Shutterstock.com

Iodine

<u>Procedure 3-Diffusion Into or Out of Dialysis Tubing</u>

1. Obtain a beaker ½ filled with distilled water and add about 3 <u>droppers-full (not drops)</u> of iodine solution (enough to make the solution golden yellow or brown). Record the color in Table 2.
2. Use a glucose test strip to test for the presence of glucose in the beaker. Dip the strip into the solution, remove immediately, and wait 30 seconds before comparing the color of the strip with the color scale on the bottle. Indicate in Table 2 if glucose is present or absent.
3. Repeat step 2 using a new test strip. Dip the strip into the beaker of the glucose/starch solution. Indicate if glucose is present or absent.
4. Obtain a pre-soaked dialysis tubing (bag) and 2 pieces of string.

5. Fold the bag over onto itself at one end and tie a **secure knot** at that end using one string. Leave about ½ inch between the string and the end of the bag.

6. Squeeze the other end of the bag between your fingers to open that end. <u>Swirl the container with the glucose/starch solution to mix</u> and use a transfer pipet to place the solution into the open end of the bag until it is about 1/2 full.

7. Squeeze the bag gently to remove excess air, fold the open end of the bag over onto itself and tie with string. The bag should not be so filled that it is tight or turgid. <u>Cut off the excess string.</u>

8. <u>Rinse the bag with tap water and blot dry with a paper towel.</u>

9. Record the color of the glucose/starch solution in the bag in Table 2.

10. Place the bag into the beaker of water and iodine.

Source: Werner Williams

11. Once the results are apparent (in about 20 minutes), record the color in both the beaker and the tube, then use a test strip to test for the presence or absence of glucose in the beaker.

Table 2. Color Records of Liquid Contents

	Color of Water in Beaker after Iodine added	Presence or absence of Glucose in Beaker Water	Presence of Glucose in Glucose/Starch Solution	Color of Glucose/Starch Solution in Dialysis Tube
Start				
After 20 minutes			Don't test	

Source: Werner Williams

1. What substance was small enough to exit the bag?

2. What substance was small enough to enter the bag?

3. What substance was too large to enter or exit the bag?

4. What are the reasons why certain molecules and ions can pass through the membrane while others do not?

Osmosis in Plant Cells

A cell will lose water when placed in a hypertonic environment. In a walled plant cell, the central vacuole will shrink, the cytoplasm will lose water by osmosis, and the plasma membrane will pull away from the cell wall. This phenomenon is known as plasmolysis. In contrast, the intake of water to a plant cell results in the cell becoming very turgid. Turgor pressure would be exerted against the cell wall. Elodea is a water plant which contain cells that have 99% H_2O and 1% salts.

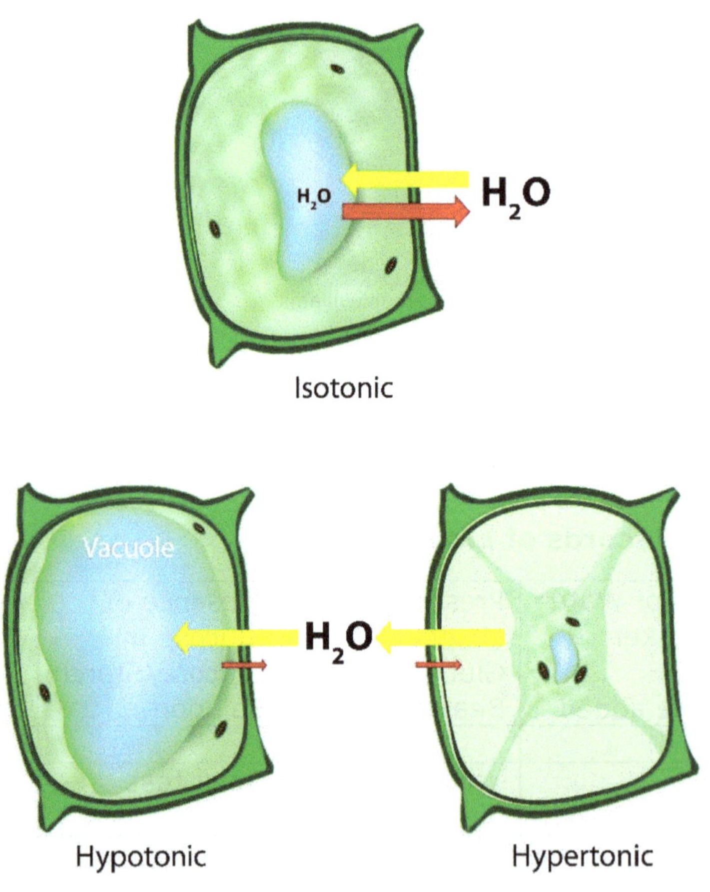

© Designua/Shutterstock.com

a. © Vladimir Arndt/Shutterstock.com   b. and c. Contributed by Kim Hakala. © Kendall Hunt Publishing Company.

a. Elodea Plant   b. Isotonic Elodea Cells   c. Plasmolyzed Elodea Cells

Procedure 4 – Demonstration of Plasmolysis

1. Observe the leaf suspended in culture water under the high power lens (40X).
2. Make a drawing of one or two representative cells. Label the cell wall, cytoplasm, and chloroplasts.
3. Observe the leaf suspended in 30% sucrose solution under the high power lens (40X).
4. Make a drawing of one or two representative cells. Label the cell wall, cytoplasm, and chloroplasts.

Plant Cells in Culture Water	Plant Cells in 30% Sucrose Solution

Source: Werner Williams

1. In which slide did the cell membrane become visible? Add a label for the cell membrane to your drawing.

2. In which slide did plasmolysis occur and why?

3. What happens when an animal cell is placed in a hypotonic environment? Contrast this to a plant cell.

Active Transport

The diffusion of molecules across a cell membrane follows the natural concentration gradient (high to low). However, the movement of molecules across a cell membrane against the natural concentration gradient (low to high) is termed active transport. This process in a living cell requires specific helper transport proteins in the membrane and requires an expenditure of energy. So, a living cell is able to admit or release substances by means of active transport.

Procedure 5 – Active Transport in Yeast Cells

Yeasts are single celled eukaryotic organisms belonging to the Kingdom Fungi.

The following solutions have been prepared. One ml of yeast solution was placed into two different test tubes. Three drops of Congo red dye were added to each test tube, labeled A and B. Test tube A was heated in a boiling water bath. Test tube B was not heated.

1. Obtain a clean microscope slide and coverslip.
2. Make a wet mount slide of test tube A.
3. Observe under low power and high power.
4. Obtain a clean microscope slide and coverslip.
5. Make a wet mount slide of test tube B.
6. Observe under low power and high power.
7. Compare the coloration of the cells in tube A to tube B. Note any differences.
8. Place the slides in the container of bleach.

Questions:

1. In which tube did the cells predominantly appear red and why?

2. In which tube did the cells predominantly appear clear and why?

3. Did you observe active transport in yeast cells? What is the evidence for this?

Lab 8: Enzymes

Reactions in the cell take place in pathways in which starting materials called **substrates(s)** are converted into **products (p).**

$$A \ + \ B \rightarrow \ C \ + \ D$$

Substrates   Products

These reactions may either combine simpler substrates in order to make complex products in a process called **anabolism** or the reactions may break down a complex substrate into simpler products in a process called **catabolism.**

Enzymes (E), are proteins that speed up the transformation from substrates to products without being changed in the process. The enzyme physically binds to the substrate and release a product(s). Fig. 1

$$E \ + \ S \ \rightarrow \ E\text{-}S \ complex \ \rightarrow \ E \ + \ P$$

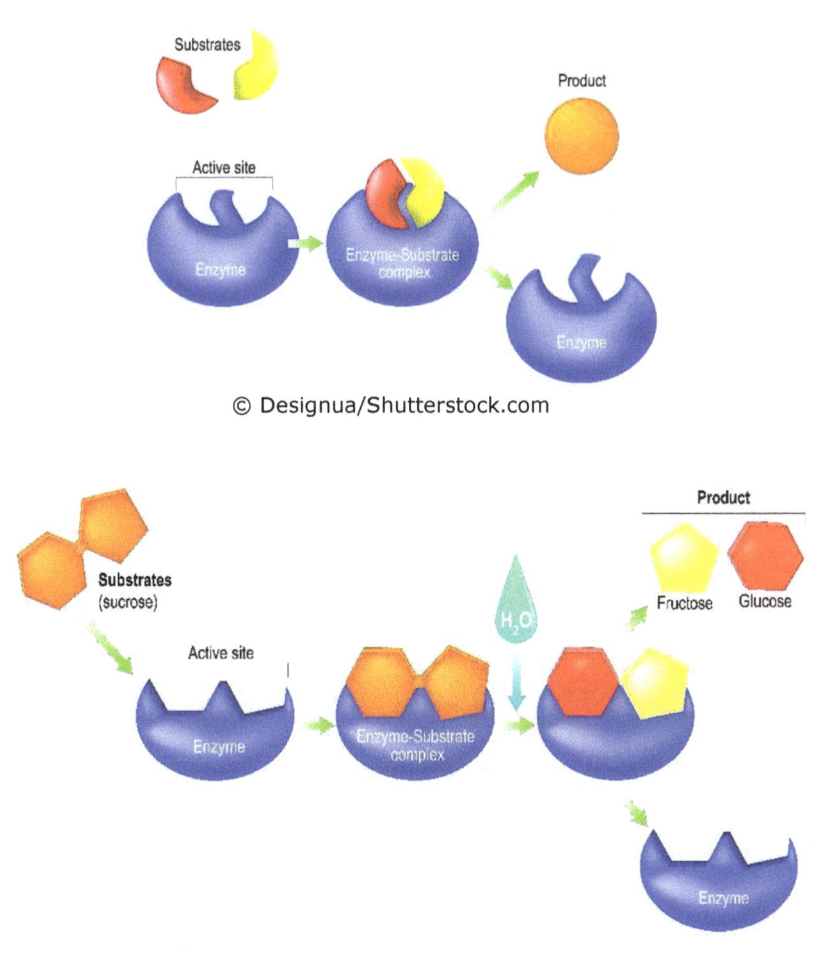

Fig. 1 Enzyme - Substrate Reactions(anabolism above & catabolism below)

The part of the enzyme that bind to the substrate is called the **active site**. That site has a shape that is unique and binds to just one substrate or type of substrate. A cell only needs a small amount of an enzyme because since the enzyme retains its original shape when the reaction is completed, it may be reused to bind to more substrates. Some enzymes can bind to thousands of substrates/second.

To function efficiently, an enzyme usually requires a closely controlled temperature and pH environment. Most enzymes need warmth and a pH that is close to neutral, but extremes in temperature or pH would generally denature most enzymes, which changes the enzyme's shape, preventing it from working. If more enzymes or substrates are added, there would be more collisions between the two, so the enzyme would work faster and more product would be made.

Glucose, fructose and galactose are simple sugars (monosaccharides) which may be linked to form two-sugar units called disaccharides. Sucrose is the disaccharide that results when glucose and fructose are linked and lactose results from linking glucose and galactose. Fig. 2

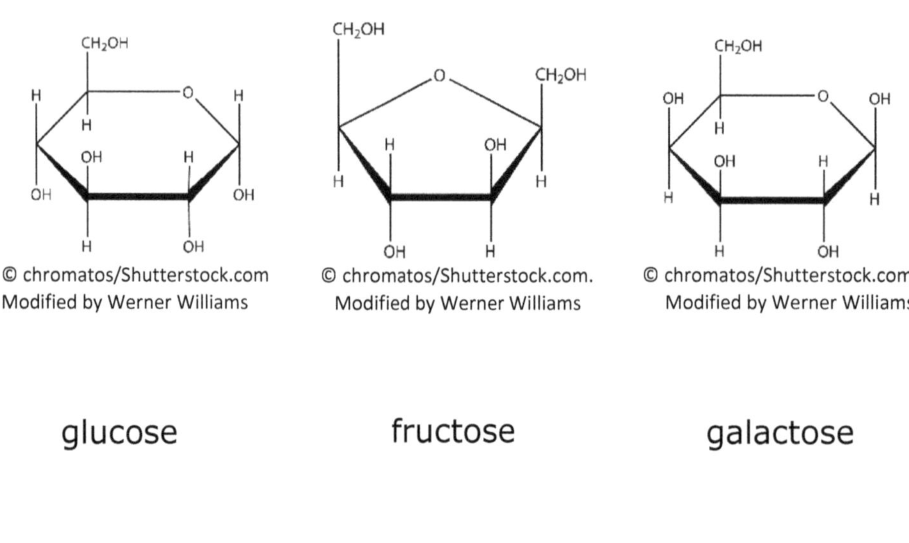

© chromatos/Shutterstock.com
Modified by Werner Williams

© chromatos/Shutterstock.com.
Modified by Werner Williams

© chromatos/Shutterstock.com.
Modified by Werner Williams

glucose   fructose   galactose

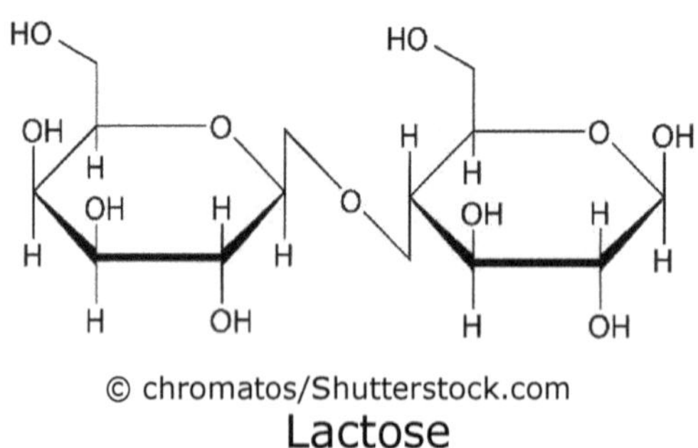

© chromatos/Shutterstock.com
Lactose

Fig. 2 Monosaccharides and Disaccharides

Lactose is found in milk (4-6% in cow's milk and 5-8% in human milk). Since lactose cannot be absorbed by the intestines, the enzyme **lactase (β-D-galactosidase)** is released which digests the lactose by breaking the glycosidic bond that links the glucose to the galactose. These sugars are then absorbed into the bloodstream. Some people have low levels of lactase which can lead to the pain and gas production of lactose intolerance when they consume dairy products. Fig. 3

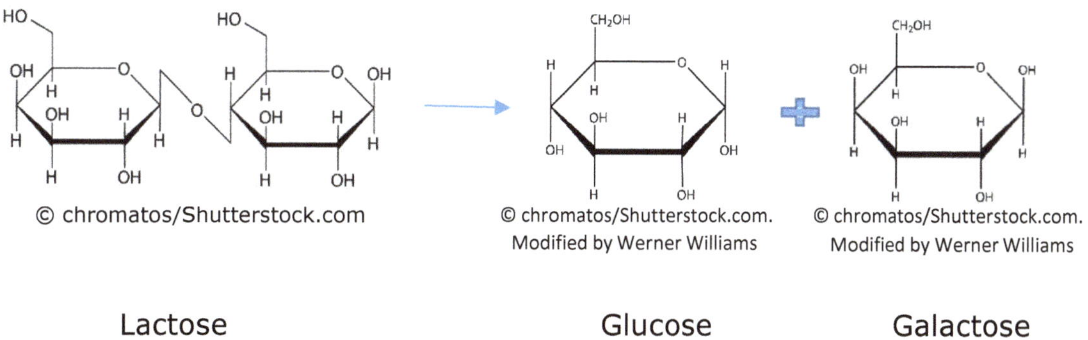

Figure 3. The Breaking Down of Lactose by Lactase

In today's lab, you will be investigating the enzyme lactase and some of the conditions (temperature, pH, concentration) that may alter its activity. Since lactase breaks down lactose into glucose and galactose, we will be using glucose test strips to detect the presence of glucose.

<u>MATERIALS</u>

Lactase	Test tube racks with test tubes
Lactose	Markers
Sucrose	Water baths
Starch	Ice bucket
Water	Incubators
Glucose test strips	Buffers (pH 3, 6, 7, 8, 11)
Ruler	

<u>Procedure 1 - The Action of Lactase</u>

1. Obtain a rack and mark 3 test tubes 1, 2 and 3.
2. Mark the tubes from the bottom at the 2 cm line and the 3 cm line.
3. Add **lactose** to the 1st line (2cm) in **tube 1**, **sucrose** to the 1st line in **tube 2** and **starch** to the 1st line in **tube 3**.
4. Add **2% lactase** to the 2nd line (3cm) in each of the three tubes.
5. Cap all tubes and then vortex to mix the tubes.
6. Allow the tubes to sit for 2 minutes.

7. Test for the presence of glucose by dipping a separate glucose test strip into each of the three tubes for at least 5 seconds. Remove the strips, wait 60 seconds and look for a color change on the square of the strip, that would indicate the presence of glucose.
8. Record your data in Table 1.

Table 1. Presence of Glucose in Various Solutions after Adding Lactase

Tube	Glucose (+ or _)
1. Lactose solution	
2. Sucrose solution	
3. Starch solution	

Source: Werner Williams

1. Which substance was digested by lactase to release glucose and galactose?

2. Why did the enzyme not react with sucrose and starch?

Procedure 2 - The Effect of Temperature on Lactase

Enzyme activity usually increases when the temperature increases. Because increasing the temperature increases the movement of the enzymes and substrates, the substrates collide more frequently with the enzyme's active site. Generally, a 10^0C rise in temperature will cause the enzyme activity to increase by two or three fold. However, very high temperature will usually denature the enzyme.

1. Obtain 6 tubes and mark them as follows: 1, 1A, 2, 2A, 3, 3A.
2. Mark each tube with your initials.
3. Mark tubes 1A, 2A and 3A at the 2cm line from the bottom.
4. Place lactose into tubes 1A, 2A and 3A to the line and cap all three tubes.
5. Mark tubes 1, 2 and 3 at the 1cm line from the bottom.
6. Place **.5% lactase** into tube 1, 2 and 3 to the line and cap all tubes.
7. Place the test tubes in the following conditions:
   Tubes 1 and 1A: ice bucket with ice (0^0C)
   Tubes 2 and 2A: 37^0C incubator (body temperature)
   Tubes 3 and 3A: boiling water bath (100^0C)

8. Let the tubes sit in the above conditions for 30 minutes.
9. **Remove the tubes. Obtain three clean transfer pipets, using one pipet for each transfer.**
10. Quickly transfer all of the contents of the <u>lactase</u> tube to the corresponding <u>lactose</u> tube (tube 1 to tube 1A, 2 to 2A etc.).
11. Cap and vortex each tube, then let them sit for 2 minutes.
12. Test for the amount of glucose by dipping a separate glucose test strip into each of the three tubes: 1, 2 and 3, for 5 seconds.
13. Referring to the test strip container, determine the relative amount of glucose in each of the three tubes (none, trace, small, moderate, large, extra- large).
14. Record your data in Table 2.

Table 2. Amount of Glucose in Solutions Incubated at Various Temperatures

Tube	Glucose (amount)
1. Incubated in ice (0^0C)	
2. Incubated at 37^0C (body temp.)	
3. Incubated at 100^0C (boiling water)	

Source: Werner Williams

1. Which of the three temperatures exhibited the most enzyme activity?

2. Explain the result when the reactants were incubated in ice?

3. Explain the result when the reactants were incubated in the boiling water bath? What happened to the lactase when it was boiled?

4. What would happen to the reaction rate if a test tube was first incubated in ice for 30 minutes, and then later incubated at 37⁰C?

<u>Procedure 3 - The Effect of pH on Lactase</u>

Each enzyme functions best at a particular pH because pH can affect the three-dimensional shape of an enzyme. Extremes in pH can denature enzymes making them inactive.

1. Obtain five test tubes and mark them 4, 6, 7, 8 and 10 according to the various pH's being tested.
2. Mark the tubes from the bottom at the 2cm, 4cm and 6cm lines.
3. Add pH 4 buffer to the first mark (2cm) of tube 4.
4. Add pH 6 buffer to the first mark of tube 6.
5. Add pH 7 buffer to the first mark of tube 7.
6. Add pH 8 buffer to the first mark of tube 8.
7. Add pH 10 buffer to the first mark of tube 10.
8. Add **2% lactase** to each of the five tubes to the second mark (4cm).
9. Cap the 5 tubes and vortex.
10. Let the tubes sit for 10 minutes.
11. Add lactose to each of the 5 tubes to the third mark (6cm) and vortex. Let the tubes sit for 2 minutes.
12. Test for the amount of glucose by dipping a separate glucose test strip into each of the five tubes for 5 seconds.
13. Referring to the test strip container, determine the relative amount of glucose in each of the three tubes (none, trace, small, moderate, large, extra- large).
14. Record your data in Table 3.

Table 3. Amount of Glucose in Various Solutions at Different pH's

Tube	Glucose (amount)
1. pH 4	
2. pH 6	
3. pH 7	
4. pH 8	
5. pH 10	

Source: Werner Williams

1. Of the pH's tested, which one was the best for lactase activity?

2. Lactase works in the small intestines. What is the pH of the human small intestines?

<u>Procedure 4 - The Effect of Lactase Concentration</u>

Because enzymes must collide with substrates, varying the enzyme concentration should have an effect on the rate of a chemical reaction.

1. Obtain three test tubes and mark them 1, 2 and 3.
2. Mark the tubes from the bottom at the 2cm and 3cm lines.
3. Add lactose to the first mark (2cm) in each of the three tubes.
4. Add **1% lactase** to tube 1 to the second mark (3cm) and vortex to mix.
5. Add **.1% lactase** to tube 2 to the second mark (3cm) and vortex to mix.
6. Add **.01% lactase** to tube 3 to the second mark (3cm) and vortex to mix.
7. Allow the tubes to sit for 2 minutes.
8. Test for the amount of glucose by dipping a separate glucose test strip into each of the three tubes for 5 seconds.
9. Referring to the test strip container, determine the relative amount of glucose in each of the three tubes (none, trace, small, moderate, large, extra- large).
10. Record your data in Table 4.

Table 4. The Amount of Glucose in Different Solutions with Varying Lactase Concentrations

Tube	Glucose (amount)
1. 1% lactase	
2. .1% lactase	
3. .01% lactase	

Source: Werner Williams

1. Which concentration showed the most enzyme activity? Why?

2. In an experiment, lactase was added to 4ml of lactose. After 1 minute, 1 ml of lactose remained. How much ml's of lactose were consumed in the reaction?_________________If we repeated the experiment and found 2.5ml of lactose remaining, would the second reaction have a higher or lower rate of reaction? __________

Lab 9: The Use of the Spectrophotometer

The ability to see colors not only adds enjoyment to our lives but it may also provide information such as when our fruits are ripe or when a storm is coming.

In this lab, you will explore how the **spectrophotometer** uses the colors of the light spectrum to determine the concentration of molecules that absorb light, found in a solution.

The visible-light spectrum, in addition to X-rays, radio waves, and infrared waves is part of the spectrum of electromagnetic radiation. Figure 1.

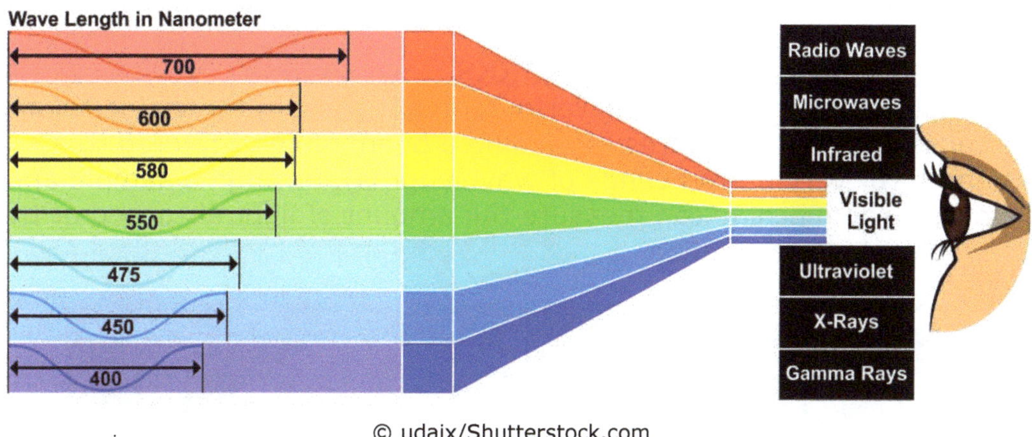

Figure 1 The Visible Light Spectrum

The different parts of this radiation differ in both wavelength and energy level, but all types travel through space in waves. The height of a wave at its crest is called the **amplitude** and the intensity or the brightness of visible light is proportional to its amplitude (the higher the amplitude, the greater its brightness). The distance from the crest of one wave to the crest of the next wave is called the **wavelength** (λ) and in the visible spectrum, the color of the light we see depends on its wavelength. Wavelength is measured in units called nanometers (1×10^{-9} m), and the wavelengths of 400-700nm make up the visible light spectrum, the part of the electromagnetic spectrum that we can see. Figure 1.

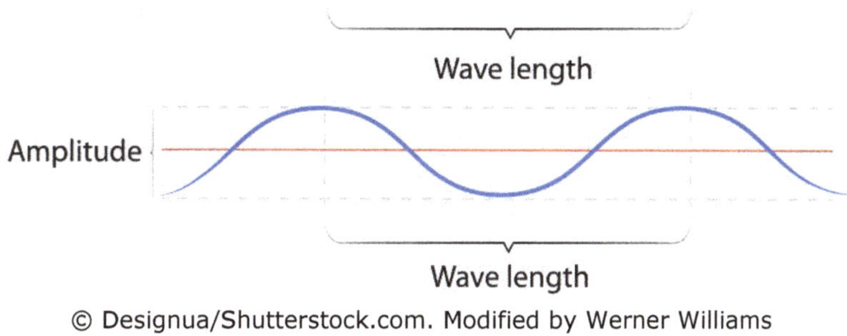

© Designua/Shutterstock.com. Modified by Werner Williams

Molecules either absorb or transmit energy in the form of electromagnetic radiation. White light (normal daylight) is made up of all the wavelengths in the visible spectrum. The color of an object is determined by how objects or molecules absorb or transmit the light that strikes them.

What we see as the color of an object or solution is determined by what wavelengths of light are "left over" to be transmitted or reflected by the object after the other wavelengths are absorbed by the object. For example, the pigment chlorophyll, present in the leaves of plants, absorbs a high percentage of the wavelengths of light in the red and violet to blue ranges. Green light is not absorbed by the chlorophyll molecules, so it is reflected from the surface of the leaf, which explains why most plants look green. A solution of chlorophyll extracted from a leaf would also be green.

HOW A SPECTROPHOTOMETER WORKS

A spectrophotometer is a very powerful tool used in both the biological and chemical sciences and operates by shining a beam of light, filtered to a specific wavelength (or very narrow range of wavelengths), through a sample and onto a light meter. The machine can measure the amount of light absorbed or transmitted by the molecules in a solution.

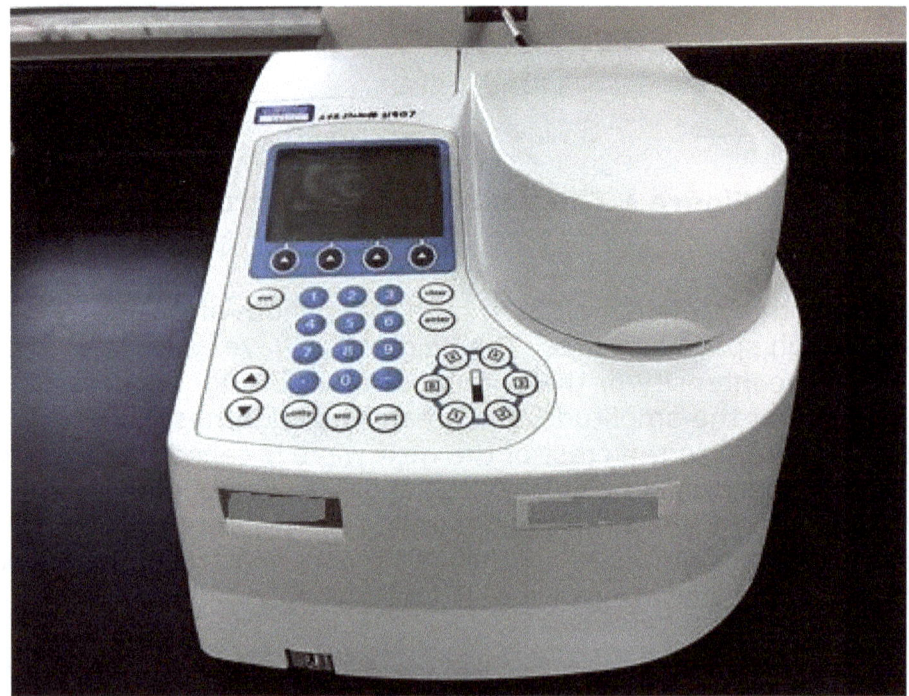

Source: Werner Williams

The light from the spectrophotometer's light source (in the case of measurements in the visible range, a simple incandescent bulb) does not consist of a single wavelength, but a continuous portion of the electromagnetic spectrum. This light is separated into specific portions of the spectrum through the use of prisms or a diffraction grating. A small portion of the separated light spectrum then passes through a narrow slit. When you adjust the wavelength on a spectrophotometer, you are changing the position of the prism or diffraction grating so that different wavelengths of light are directed at the slit. This small band of light then passes through the tube containing the sample, and the light is detected by a photocell and measured to yield the transmittance or absorbance value for the sample. Figure 2.

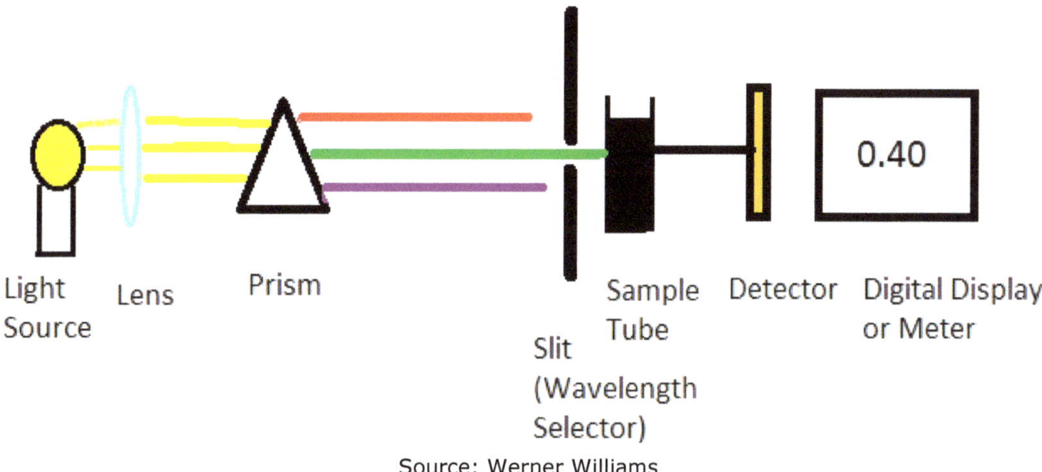

Source: Werner Williams

Figure 2 How a Spectrophotometer Works

A given substance will not absorb all wavelengths equally. Because different substances absorb light at different wavelengths, a spectrophotometer can be used to identify compounds by analyzing the pattern of wavelengths absorbed by that substance. Additionally, the amount of light absorbed is directly proportional to the concentration of the absorbing substances in that sample, so we can use the machine to determine the concentration of a solute. Finally, because particles in a mixture will scatter light (thus preventing it from reaching the light detector), spectrophotometers may also be used to estimate the number of cells in a mixture.

Spectrophotometers that use ultraviolet or visible light are the types most often used to study biological structures or reactions. The investigator can select the specific wavelength of light that will be maximally absorbed by the particular substance that is in the solution.

Regardless of whether a solution is colored or colorless, the wavelengths of light that are absorbed and transmitted by a substance are distinctive and are called its <u>absorbance spectrum</u>. A solution like crystal violet absorbs light at specific wavelengths and reflects light at other wavelengths. If a solution absorbs green and red light, but not blue light (it reflects blue), it will look blue. When we look at the absorption spectrum of that blue solution, it should show low absorption of wavelengths in the blue part of the spectrum. Figure 3.

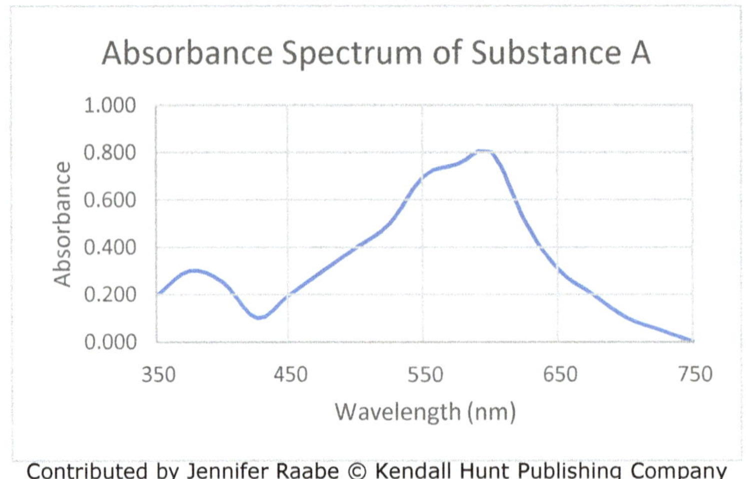

Contributed by Jennifer Raabe © Kendall Hunt Publishing Company

Figure 3 An example of an absorbance spectrum

To insure that spectrophotometer readings indicate only the concentration of the solute we wish to measure, we must first get a reading using a **blank**. Ideally, the blank should contain everything found in the sample solution except the substance you are trying to analyze or measure. When we measure the blank tube, the instrument looks at the molecules in that tube and reads the tube as 100% transmittance or zero absorbance. When the machine then measures the sample tube, the only substance absorbing light (if you are measuring absorbance) or transmitting light (if you are measuring % transmittance) will be the actual molecule that we are attempting to measure.

Let's look at an example of how a blank would be prepared. A 1 ml sample of substance X is mixed with 5 ml of water and 1 ml of a colored reagent to give a total volume of 7 ml in a sample tube. A blank is prepared by mixing 5 ml of water with an additional 1 ml of water (as a substitute for substance X) and 1 ml of the colored reagent. <u>The final volume in the blank tube must also be 7 ml (the same volume as in the sample tube)</u>.

<u>Procedure 1 – Determining the Absorption Spectrum of Crystal Violet</u>

In this procedure, you will determine the absorption spectrum of crystal violet.

1. Obtain 2 tubes and 1 cap.
2. Label one of the tubes "B" near the top edge of the tube for the blank, and place 5ml of distilled water into the tube.
3. Place 2.5 ml of distilled water into the second tube along with 2.5ml of crystal violet. This is your crystal violet sample tube.
4. Cap and vortex the crystal violet sample tube.
5. Press the "**Basic ATC**" button (lower right of the display screen).
6. To set the wavelength, press the "**Set nm**" button.
7. Enter **410** in the numeric pad then press **"Set nm"**.

8. Wipe the outside of the tubes that will be placed in the machine with Kim wipes.
9. Open the cover on the top right of the machine and place the "blank" tube in the **"B"** tube slot.
10. Place the other tube into slot #1.
11. Close the cover. Find the section of the control panel that has the circle with the B and tubes 1,2,3,4 and 5 (*Circle tubes section*). Press the **"B"** button.
12. Press the "**Measure Blank**" button (lower left of the display screen). Wait until the 0.00A appears.
13. Since you will be measuring absorbance (A), make sure you see an **A** in the display. If you see a **%T** (transmittance) or **C** (concentration), press the _ "**Change Mode**" button (lower right of the display screen) until an **A** appears.
14. Press button **"1"** in the *circle tubes section* to measure tube #1. Take the reading in absorbance (A) ["Tube 1" appears on the upper right of the display screen].
15. Set the wavelength at intervals of 20nm until you reach 650nm. <u>Set the blank at each wavelength and measure the absorbance of the sample tube.</u> **At each higher wavelength repeat steps # 6, 7 and 11-14 of these instructions.**
16. Record your results in **Table 1**
17. **Remove your tubes and press "clear" or "esc" (escape).**

Table 1

Wavelength(nm)	Absorbance	Wavelength(nm)	Absorbance
410		550	
430		570	
450		590	
470		610	
490		630	
510		650	
530			

Source: Werner Williams

Graph the absorption spectrum of crystal violet. The X-axis should be wavelength and the Y-axis should be absorbance.

At what wavelengths is crystal violet maximally absorbed?
____

At what wavelengths is crystal violet minimally absorbed?
____

Do your results show low absorbance in the blue range?
____

<u>**Procedure 2- Examining the Relationship Between Absorbance and Transmittance.**</u>

To observe the relationship between absorbance and % transmittance, prepare different concentrations of crystal violet.
1. Obtain 4 tubes and 4 caps. Label them at the top with a marker: B, 1, 2, 3.
2. Place 6 ml of water in the B tube. This is your blank.
3. In tube #1, place 1 ml of crystal violet and 5 ml of water.
4. In tube #2, place 3 ml of crystal violet and 3 ml of water.
5. In tube #3, place 5 ml of crystal violet and 1 ml of water.

6. Cap, and then vortex tubes 1-3.
7. Press the "**Basic ATC**" button (lower right of the display screen).
8. To set the wavelength, press the "**Set nm**" button.
9. Enter **590** in the numeric pad then press **"Set nm"**.
10. Wipe the outside of the 4 tubes with Kim wipes.
11. Open the cover on the top right of the machine and place the "blank" tube in the **"B"** tube slot.
12. Place the other tubes into their respective slots. **Tube # 1 goes into slot #1; tube # 2 goes into slot #2 ...** and so on.
13. Close the cover. Find the section of the control panel that has the circle with the B and tubes 1,2,3,4 and 5 (*Circle tubes section*). Press the **"B"** button.
14. Press the "**Measure Blank**" button (lower left of the display screen). Wait until the 0.00A or 100%T appears.
15. **Measure the absorbance (A) first.** Press the "**Change Mode**" button (lower right of the display screen) until an A appears on the display (**%T** stands for transmittance and **C** stands for concentration).
16. Press button **"1"** in the *circle tubes section* to measure tube #1. Take the reading in absorbance (A).("Tube 1" appears on the upper right of the display screen).
17. Press button "**2**" and so on to measure the absorbance of the other tubes.
18. Press the **"B"** button in the *circle tubes section*.
19. Press the "**Measure Blank**" button (lower left of the display screen). Wait until the 100%T appears. (You may press the "**Change Mode**" button [lower right of the display screen] until a %T appears on the display).
20. You are now ready to measure the %T of the tubes. Press button **"1"** in the *circle tubes section* to measure tube #1. Take the reading in %T. ("Tube 1" appears on the upper right of the display screen).
21. Press button "**2**" and so on to measure the %T of the other tubes.
22. Record **both** the absorbance and the %T for tubes 1-3 in table 2.
23. **Remove your tubes and press "clear" or "esc" (escape).**

Table 2   Absorbance and %T of Various Concentrations of Crystal Violet

Tubes	Absorbance	% Transmittance
1 (1 ml crystal violet)		
2 (3 ml crystal violet)		
3 (5 ml crystal violet)		

Source: Werner Williams

Does the absorbance increase or decrease with the increasing intensity of the color?
____

Does the %Transmission increase or decrease with the increasing intensity of the color? ____

<u>Procedure 3- Determining Transmittance and Absorbance</u>

Coomassie blue binds to proteins like albumin and forms a blue complex. The darker the color, the more protein is present. The intensity of the color can be accurately measured by determining the amount of light that is absorbed by the solution. You will be using a **transfer pipet** to measure the water and the drops of albumin.

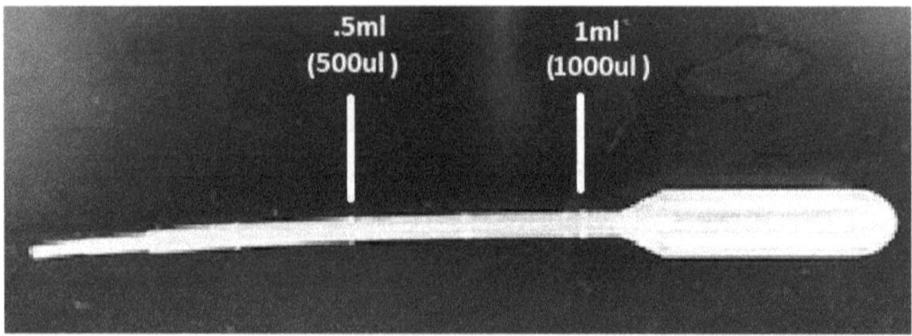

Source: Werner Williams

1. Obtain two tubes and label one "S" (for the sample) and the other "B" (for the blank).
2. Prepare a sample tube by placing 10 drops of albumin solution into tube S and then add 1 ml of distilled water using the provided transfer pipets (See measurements on the picture of the transfer pipet above).
3. Prepare a blank tube by placing 10 drops of distilled water into tube B and then add <u>an additional 1 ml of distilled water </u>by using the <u>same transfer pipet.</u>
4. Add 5 ml of Coomassie blue to both tubes using the provided pipet.
5. Cap and vortex the tubes, and allow them to stand for 5 minutes.
6. Press the "**Basic ATC**" button (lower right of the display screen).
7. Press the "**Set nm**" button.
8. Enter **595** in the numeric pad then press **"Set nm"**.
9. Wipe the outside of the tubes with Kim wipes.
10. Place the "blank" tube in the **"B"** tube slot and the "sample" tube in slot #1.
11. Close the cover. Press the **"B"** button in the *Circle tubes section*.
12. Press the "**Measure Blank**" button (lower left of the display screen). Wait until 0.00A appears. If it doesn't appear, press the "**Change Mode**" button (lower right of the display screen) until an A appears on the display.
13. Press button **"1"** in the *circle tubes section* to measure the absorbance of the sample tube.
14. Press the **"B"** button in the *Circle tubes section*.
15. Press the "**Measure Blank**" button (lower left of the display screen). Wait until the 100%T appears. If it doesn't appear, press the **"Change Mode"** button until %T appears.
16. Press button **"1"** in the *circle tubes section* to measure the % transmittance.
17. **Remove your tubes and press "clear" or "esc" (escape).**
    Record your data: absorbance ____________________________________
    transmittance ____

<u>Procedure 4- Exploring the Relationship Between Absorbance and Concentration</u>.

Absorbance is a function of concentration. A higher absorbance results from a higher concentration. By comparing the absorbance of a solution containing an unknown amount of protein to a **standard curve**, which is a graph of the absorbance values plotted from a series of samples of known concentrations of the same material, the concentration of protein in an "unknown" sample can be determined. Figure 4.

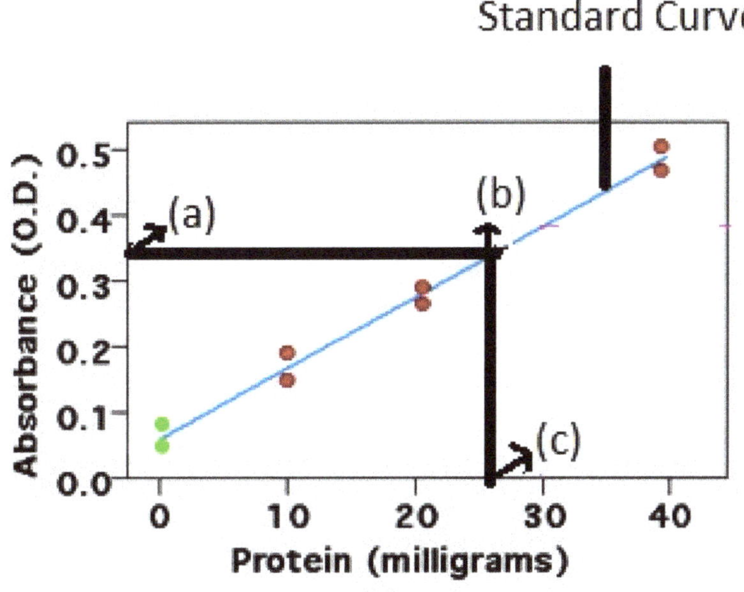

Source: Werner Williams

Figure 4- The absorbance measurements of known concentrations of a protein molecule are used to generate a standard curve. In this example, the "curve" is linear. An absorbance reading (a) is obtained for a solution containing an unknown amount of protein. By drawing a straight line from (a) to the curve at (b), and dropping a line from (b) to (c), the concentration of the protein can be determined. The Absorbance (O.D) (Y-axis) indicates absorbance at 595 nm.

1. Obtain a tube from your instructor marked BSA. It contains 240 µg/ml of bovine serum albumin (BSA), a purified form of albumin. To develop a standard curve for BSA, your group should prepare at least five dilutions.
2. Label 5 tubes at the top with the numbers 1, 2, 3, 4, and 5. Label a sixth tube "B" for the blank.
3. Prepare dilutions as follows: Use a <u>blue micropipet </u>and a tip to add 0.5 ml of distilled water to each of the tubes (#'s 1,2,3,4, 5 and to the blank).
4. Using the same micropipette tip, take 0.5 ml from the BSA tube and add that to tube #1. This tube now has a protein conc. of 120 µg/ml. <u>Cap and vortex the tube</u>.
5. Take 0.5 ml of solution from test tube # 1 and add it to test tube #2. <u>Cap and vortex the tube.</u>

6. Take 0.5 ml of solution from test tube #2 and add it to test tube #3. <u>Cap and vortex the tube.</u>
7. Take 0.5 ml of solution from test tube #3 and add it to test tube #4. <u>Cap and vortex the tube.</u>
8. Take 0.5 ml of solution from test tube #4 and add it to test tube #5. <u>Cap and vortex the tube.</u>
9. Take 0.5 ml from test tube #5, discard it into a sink **and also discard the micropipette tip in the trash**.
10. Obtain an "unknown" tube from your instructor which contains 0.5 ml of a concentration of BSA. Your unknown tube # is_______________
    You should now have a series of five test tubes containing protein concentrations of 120, 60, 30, 15, and 7.5 µg/ml; a blank containing only water; and a tube containing an unknown concentration of protein.
11. Add 5 ml of Coomassie blue using a <u>green micropipet </u>to each tube (#'s 1-5, the blank and the unknown). Cap all tubes and vortex. <u>Wait 5 minutes before proceeding</u>.
12. Press the "**Basic ATC**" button (lower right of the display screen).
13. Press the "**Set nm**" button, enter **595** in the numeric pad then press **"Set nm"**.
14. Wipe the outside of each of the 7 tubes with Kim wipes.
15. Place the "blank" tube in the **"B"** tube slot and the other tubes into their respective slots. **Tube # 1 goes into slot #1; tube # 2 goes into slot #2 ... and so on.**
16. Close the cover. Press the **"B"** button in the *Circle tubes section* and then press the "**Measure Blank**" button (lower left of the display screen) and wait until the 0.00A appears. If it does not appear, press the "**Change Mode**" button (lower right of the display screen) until an A appears on the display(**%T** stands for transmittance and **C** stands for concentration).
17. Press button **"1"** in the *circle tubes section* to measure tube #1. Take the reading in absorbance (A).
18. Press button "**2**" and so on to measure the absorbance of the other tubes.
19. Remove tube **1**, replace it with the unknown, close the cover and press **1** in the circle tubes section to measure the absorbance of the unknown.
20. Record your data below in Table 3, **remove your tubes and press "clear" or "esc" (escape)**

Table 3 Protein Concentration vs. Absorbance Values

	Protein Conc. (µg/ml)	Absorbance (595nm)
Tube 1	120	
Tube 2	60	
Tube 3	30	
Tube 4	15	
Tube 5	7.5	
Unknown		

Source: Werner Williams

Graph your results. Label the X-axis "concentration" and the Y axis "absorbance". Make the graph as large as possible. **Plot the absorbance data for the solutions of known concentrations (tubes 1-5)**. This is your standard curve. Keep in mind that a standard curve can be used as a standard only when the known and unknown concentrations have been prepared according to the same procedure. For example, for this standard curve to be useful, the tubes with the known concentrations and the "unknown" tube must be prepared by using 0.5 ml of the sample and 5 ml of the colored reagent.

Determine the concentration of the unknown._______________________µg/ml BSA.
Record this value in Table 3.

1. You have a solution that appears orange. What color light is being transmitted?

2. How is absorbance related to transmittance of light through a solution?

3. What is the purpose of a "blank" sample?

4. You are making a standard to be used to measure protein concentration. You have added 2 ml of protein, 3 ml of water, and 4 ml of coomassie blue. How do you make a blank for this experiment?

5. In procedure 4, you discarded 0.5 ml from tube #5 (see step #9 in the procedure). Why was that done?

Lab 10: Cell Respiration

Microorganisms and Cell Respiration

Yeasts are microscopic fungi that perform cellular respiration using similar pathways to those found in larger plants and animals. Methylene blue can be used to determine if cell respiration is occurring in yeast. Figure 1

Aerobic respiration releases hydrogen ions and electrons which are absorbed by the methylene blue, gradually turning the dye colorless. Clear cells are alive and dead cells will remain blue because they are not undergoing respiration.

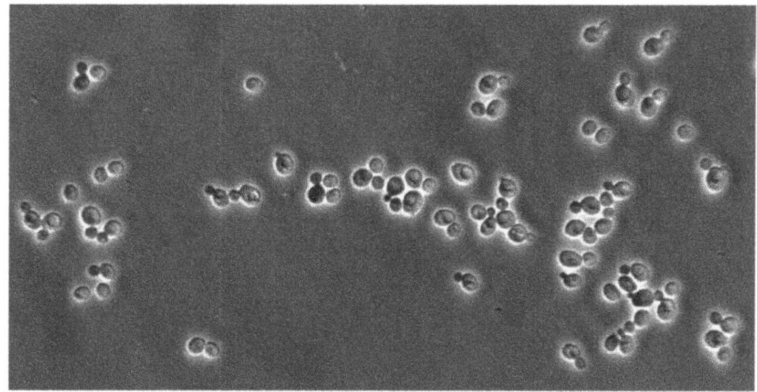

© Andre Nantel/Shutterstock.com

Figure 1 Microscopic View of Yeast Cells

Procedure 1

1. Place a drop of yeast suspension on a clean microscope slide. Add one small drop of methylene blue to the yeast and <u>wait 5 minutes</u>.
2. Place a cover-slip over the yeast. Use the low power lens (10X) and then the high power lens (40X) to observe the cells that are colorless and the ones that are blue. Place the slide in the beaker of bleach when finished.

Questions

1. Is cell respiration taking place in all cells on your slide? _________________
2. Estimate the percentage of yeast cells that are not undergoing cell respiration?
3. Why isn't cell respiration occurring in all the cells?_____________________

____

<u>**Comparison of Classroom Air with Exhaled Air**</u>

This procedure uses <u>limewater</u> (a Calcium carbonate solution) as an indicator solution. <u>It turns cloudy when CO_2 is added to it</u>. Would you expect room air to cause limewater to turn cloudy?___________Would you expect exhaled air to cause limewater to turn cloudy? _______________________________

<u>Procedure 2</u>

1. Label two tubes #1 and #2, and mark them at the 5 cm line.
2. Using the squeeze bottle, add limewater to the mark on each of the tubes.
3. Using the pipet with the rubber bulb, continuously bubble air into tube #1 for one minute. Record result.
4. Using the drinking straw, **very gently** bubble your exhaled air into tube #2 for one minute. Record result.
   (Be careful not to suck the solution into your mouth!)
5. Discard the straw.

Result

1. Observation of tube 1.____

   ____

2. Observation of tube 2.____

   ____

3. What gas was bubbled into tube #2? _______________________________
4. What process (glycolysis, Krebs cycle, electron transport chain) produced the gas?

   ____

5. Name the part of the cell where glycolysis takes place.___________________
6. Name the part of the cell in which aerobic respiration pathways occur.

   ____

7. Name the part of the cell in which anaerobic respiration pathways occur.

   ____

<u>**Anaerobic Respiration**</u>

Cells are able to produce energy in the absence of oxygen. This process is called anaerobic respiration (fermentation) and occurs in the cytoplasm. Carbon dioxide may be produced as well. Figure 2

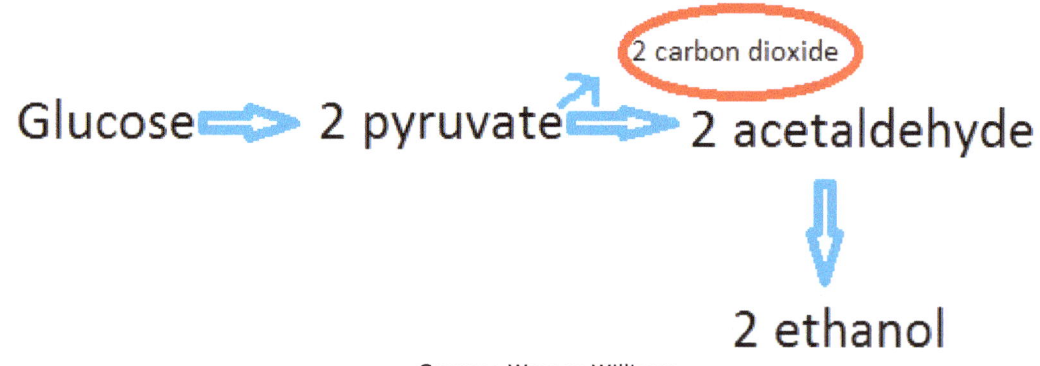

Source: Werner Williams

Figure 2 Alcoholic Fermentation in Yeast Producing CO$_2$

We will observe 2 flasks, only one of which is undergoing anaerobic respiration. Figure 3

Flask A   Flask B

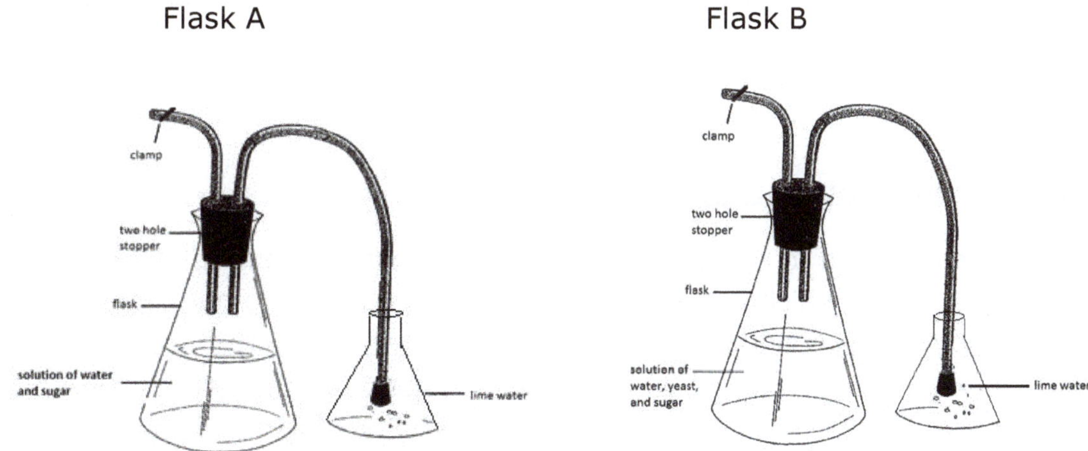

Contributed by Kim Hakala. © Kendall Hunt Publishing Company.

Figure 3 Demonstration of Anaerobic

Respiration

<u>Procedure 3</u>

1. Observe 2 flasks, A and B, on warming plates. Both flasks contain 100 ml H_2O and 2 g of glucose. Flask A contains no yeast, while flask B contains 4g of yeast.
2. Observe any change to the limewater in the flask that is connected to both flask A and flask B. Does it turn cloudy after about 45 minutes?

____

____

Result

1. Color of limewater connected to flask A after 45 minutes? _________________
2. Color of limewater connected to flask B after 45 minutes? _________________
3. Presence of bubbles in collecting flask? ___________________________________
4. What gas was being released that formed bubbles? _________________________
5. A yeast cell undergoing fermentation produced 8 ATP molecules. How many glucose molecules were used?____________________________________Explain your answer. (See 1st paragraph "aerobic respiration" –next page). _______________

____

____

6. If the same cell was undergoing aerobic cellular respiration instead of fermentation, but using the same number of glucose molecules, how many ATPs would be produced?_____________Explain.

____

____

<u>Respiration in Plant Embryos</u>

Several tests are used to check if seeds are viable before planting. One test uses the dye <u>tetrazolium, which is colorless when oxidized but becomes reddish when reduced. The test is dependent on the seed's functioning electron transport chain. If the cells in a seed are viable when tetrazolium is added, then the cytochromes in the electron transport chain will lose electrons to tetrazolium. The tetrazolium becomes reduced and turns red</u>. If the seed is dead, then its electron transport chain will not be functioning and electrons will not be sent to tetrazolium. The dye will remain colorless.

<u>Procedure 4</u>

1. Two corn seeds have been placed in separate petri plates after soaking in water overnight. Corn seed B was boiled.
2. A few drops of tetrazolium were placed in contact with the corn in the plates.
3. Observe the corn to see whether A or B are red. (**Caution- Tetrazolium is a poison. Avoid contact with the skin. Wash immediately if it touches your skin).**
4. Which seed (A or B) is actively respiring?_______________________________

5. Cyanide is a poison that affects the electron transport system by binding to the cytochromes, which prevents the electrons from flowing completely through the chain of carriers. If seeds were treated with cyanide, then what would they show in the tetrazolium test? _______________________

____

Aerobic Respiration

Most organisms including humans produce most of their ATPs by using aerobic respiration rather than fermentation. Aerobic respiration begins with glucose, which is also used in fermentation. However, aerobic respiration extracts more energy. A single glucose molecule can produce as many as 30-32 ATPs in aerobic respiration (depending on whether the cell is eukaryotic or prokaryotic), compared to only 2 ATPs per glucose in fermentation.

Like fermentation, aerobic respiration begins with <u>glycolysis</u>, but in aerobic respiration, pyruvate, a product of glycolysis enters the second set of reactions called the <u>Kreb's (or citric acid) cycle</u>. Some of the reactions in the Kreb's cycle produce $NADH_2$ and $FADH_2$ which release protons and electrons, which then enter the <u>electron transport system.</u> Most of the ATPs made in aerobic respiration are made by the electron transport system.

Glycolysis takes place in the cytoplasm, while the Kreb's cycle and the electron transport chain take place in the mitochondria, which are found in all eukaryotic cells. You will use a suspension of ground-up lima beans to study aerobic respiration. The suspension contains mitochondria. Under our experimental conditions, mitochondria will continue to carry out aerobic respiration as if they were found in intact cells. Sucrose has been added to the suspension as a source of glucose for respiration.

The rate of aerobic respiration can be measured by studying the activity level of the enzyme succinic dehydrogenase in the Kreb's cycle. The enzyme speeds up a reaction that converts succinate (the substrate) into fumarate. Figures 4 and 5

(In Figure 5, note that succinate loses 2 H$^+$ [in red] to FAD, which becomes FADH$_2$, & the succinate is converted to fumarate which retains some of its H's [in green]).

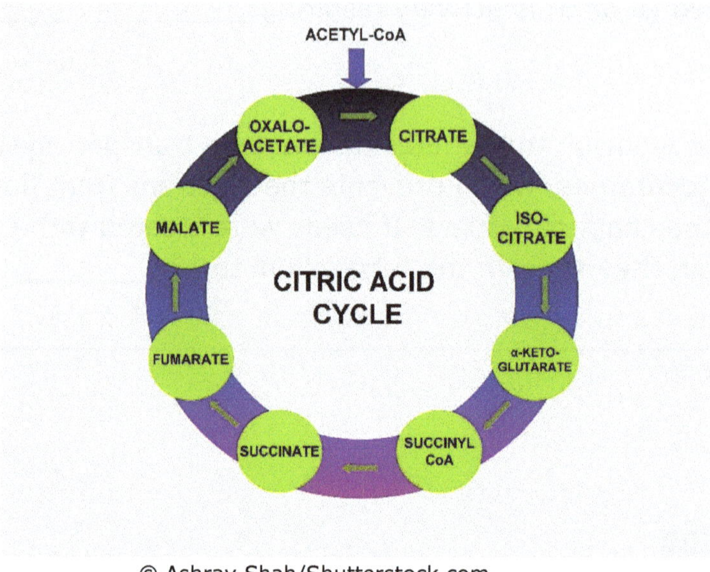

© Ashray Shah/Shutterstock.com

Figure 4 The Reaction of Succinate to Fumarate in the Kreb's Cycle

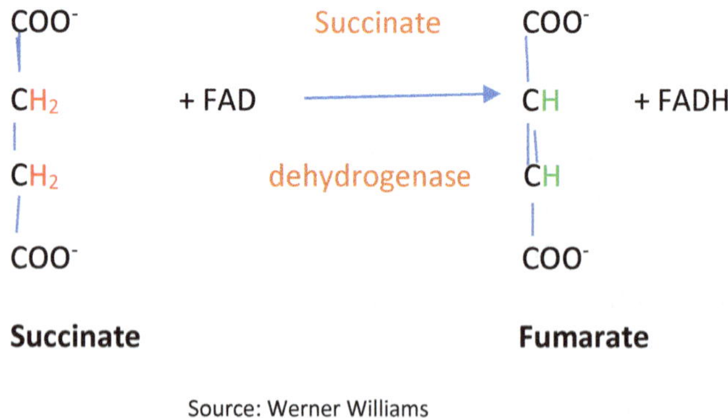

Source: Werner Williams

Figure 5 Conversion of Succinate to Fumarate

As the reaction proceeds, protons and electrons are released from the succinate, and this reaction is catalyzed by the enzyme succinic dehydrogenase using the co-enzyme FAD, which is reduced to $FADH_2$. In the living mitochondrion, these protons and electrons are carried to the electron transport system and used to make ATP. You will add a blue solution, DCPIP (di-chlorophenol-indophenol) to the reaction mixture. DCPIP intercepts the protons and electrons before they are received by FAD and sent to the electron transport system. When a DCPIP molecule picks up a proton and electron, its blue color disappears, and it gradually becomes colorless. We can use this change of color as a measure of the enzymatic rate, which is also a measure of the rate of aerobic respiration. As succinate is converted to fumarate, and as aerobic respiration proceeds, the solution in the test tube gradually turns from blue to colorless. The reaction is summarized below in Figure 6:

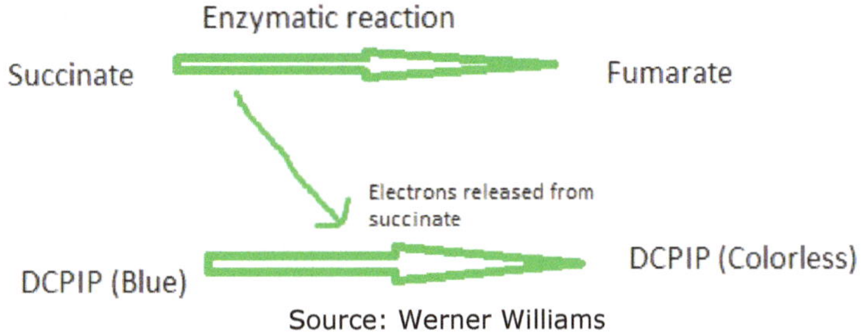

Figure 6 The Change of DCPIP from Blue to Colorless

Malonate should stop the reaction from occurring because it acts as a competitive inhibitor by also binding to succinic dehydrogenase. Figure 7.

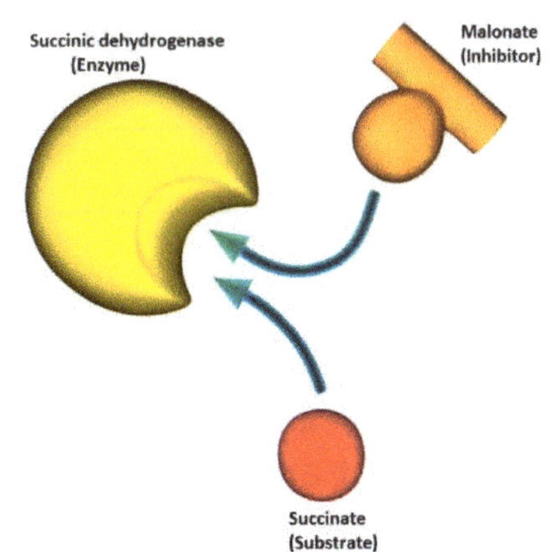

© sciencepics/Shutterstock.com. Modified by Werner Williams

Figure 7 Succinate (Substrate) and Malonate (Inhibitor) Compete

for the same enzyme.

We will be using a spectrophotometer to measure the change in color. Your data will show how much light has been transmitted by the sample. <u>The more intense the color of the sample, the less light it transmits. The percent transmittance measures how much of the blue color has disappeared. Since the blue DCPIP becomes colorless as aerobic respiration proceeds, the transmittance reading becomes higher, as more mitochondrial activity is taking place.</u>

Before beginning the procedure, look at the table below to see how the experiment will be set up.

Materials	Blank (volume in ml)	Tube 1 (volume in ml)	Tube 2 (volume in ml)	Tube 3 (volume in ml)	Tube 4 (volume in ml)
Buffer	4.6	4.4	4.3	4.1	4.1
DCPIP	0	0.3	0.3	0.3	0.3
Mitochondrial suspension	0.3	0.3	0.3	0.3	0.3
Succinate	0.1	0	0.1	0.3	0.1
Malonate	0	0	0	0	0.2

Source: Werner Williams

What is the role of each of the following?

1. Lima been extract____

____

2. Succinate____

____

3. DCPIP____

____

4. Buffer____

____

5. Malonate____

6. Why should there be a different amount of buffer in each tube?____________

____

7. What hypothesis is being tested?____

____

8. Predict the outcome of the experiment in terms of your hypothesis._______

____

Procedure 5

The mitochondrial suspension is slightly cloudy. Even though it is colorless, transmittance of light through the sample will already be less than 100% before any DCPIP is added. The blank will have to contain mitochondrial suspension but no DCPIP.

1. Obtain 5 tubes and label them 1, 2, 3, 4 and B for the blank. Measure the appropriate amounts of the solutions into tubes B, 1, 2, 3 and 4 according to the table below, **but do not add the succinate as yet**. You may need to adjust pipet settings for correct measurements.

Materials	Blank (volume in ml)	Tube 1 (volume in ml)	Tube 2 (volume in ml)	Tube 3 (volume in ml)	Tube 4 (volume in ml)
Buffer	4.6	4.4	4.3	4.1	4.1
DCPIP	0	0.3	0.3	0.3	0.3
Mitochondrial suspension	0.3	0.3	0.3	0.3	0.3
Succinate	0.1	0	0.1	0.3	0.1
Malonate	0	0	0	0	0.2

Source: Werner Williams

2. Press the "**Basic ATC**" button (lower right of the display screen).
3. Press the "**Set nm**" button.
4. Enter the wavelength (**600**) in the numeric pad then press "**Set nm.**"
5. As quickly as possible, add the appropriate amounts of succinate according to the table above to the **tubes B, 2, 3 and 4**.
6. Cap & vortex all tubes. **Check each tube to make sure the blue color is well mixed throughout each tube.**
7. Wipe the outside of the tubes that will be placed in the machine with Kim Wipes.
8. Open the cover on the top right of the machine and place the "blank" tube in the "**B**" tube slot.
9. Place tube 1 into tube slot #1, tube 2 into slot #2, tube 3 into slot #3 and tube 4 into slot #4.
10. Close the cover. Find the section of the control panel that has the circle with the B and tubes 1, 2, 3, 4 and 5 (*Circle tubes section*). Press the "**B**" button.

11. Press the "**Measure Blank**" button (lower left of the display screen).
12. "100%T" should appear on the display screen (99% is close enough).
13. Since you will be measuring transmittance (%T), make sure you see a **%T** in the display. If you see an A (absorbance) or **C** (concentration), press the "**Change Mode**" button (lower right of the display screen) until a **%T**.
14. Press button "**1**" in the *circle tubes section* to measure tube 1. ("Tube 1" appears on the upper right of the display screen). This %T value will for at 0 minutes. Record that value in the table below.
15. Press button "**2**" in the *circle tubes section* to measure tube 2 and also record this value at 0 minutes..
16. Repeat step #15 above for tubes 3 and 4 (press buttons "**3**" and then "**4**") and record their values as well. **Remove your tubes**.
17. Wait 10 minutes, take the readings again (repeat steps #6-16) in %T, and record the values. **Remove your tubes**.
18. Wait another 10 minutes and take the readings again (steps #6 to 16) and record the values.
19. **Remove your tubes and press "clear" or "esc" (escape).**

Tube	0 minutes	10 minutes	20 minutes
1(mitochondrial suspension-MS)			
2(MS +0.1 ml succinate)			
3(MS+0.3 succinate)			
4(MS+0.2 malonate)			

Source: Werner Williams

Plot the % transmittance for each tube against time *(%T on the Y axis and Time on the X-axis)*. This will show the reaction rate for aerobic respiration. The steeper the slope of the line, the faster the reaction rate. Be sure to label the axes completely.

1. Was the hypothesis proven false or supported by the results? (Use data to support your answer.

____

____

____

2. Which reaction tube was the control?_________________________________

3. Why should the succinate be added to the reaction tubes last?___________

____

____

4. What are some independent variables that could affect aerobic respiration?

____

____

<u>Lab 11: Cell Division</u>

New cells are created when older cells divide. An important part of cell division is the replication of the DNA molecule in the **parent cell** and its distribution to each of the **daughter cells**. DNA contains the hereditary material, controls cellular functions and is located in the chromosomes.

A distinct number of chromosomes are found in the body cells of each biological species. Cabbage has 18 chromosomes, a snail has 24 and humans have 46. Chromosomes are usually found in pairs with both containing similar hereditary information, so each pair is said to be made up of two **homologous chromosomes**. A cell is said to be **diploid** (designated **2n**) if its chromosomes are found in pairs and is said to be **haploid** (designated **n**) if only one set of chromosomes is present.

A chromosome before it replicates consists of two arms joined together at a point called the **centromere**. Each chromosome then replicates so each is then made up of two identical **sister chromatids** joined at the centromere. Each chromatid is made up of a single DNA with proteins and is identical to the original chromosome before it replicated. When the sister chromatids are paired, they are considered to be a single chromosome, but after they separate, each chromatid becomes a complete independent chromosome called a daughter chromosome. Figure 1.

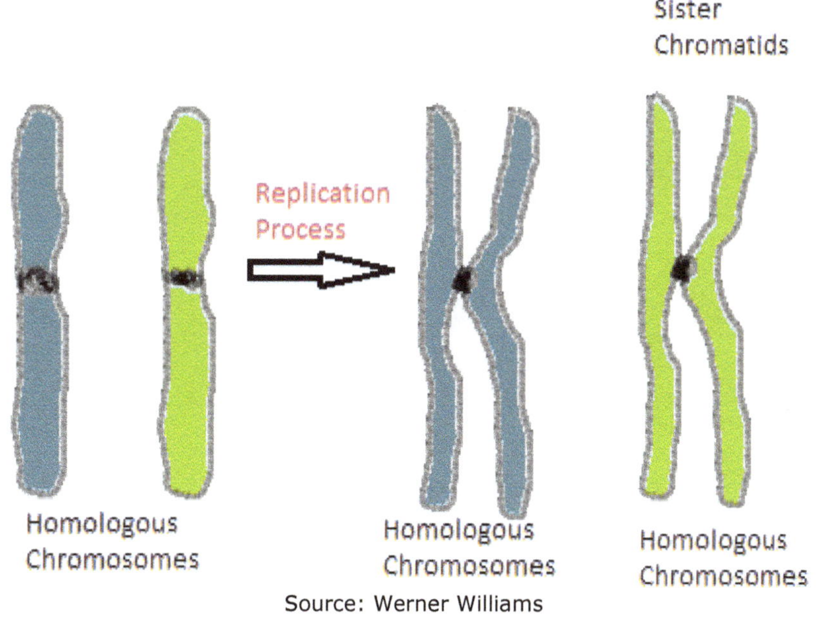

Figure 1 Chromosomes and Chromatids

In eukaryotic cells, **mitosis and meiosis** are the two processes involved in separating and distributing the chromosomes from the parent cell to the resulting daughter cells. **Cytokinesis** is the division of the cytoplasm, which results in the formation of daughter cells.

Somatic (body) cells have the same number of chromosomes as the parent cell and are produced by mitosis and cytokinesis. **Gametes (eggs and sperm)** have half the number of chromosomes as the parent cell and are produced by meiosis and cytokinesis.

The Cell Cycle

A cell passes through distinct stages during its life span called the **cell cycle**. The two major stages are **interphase** and the **M (mitosis)** stages. In interphase, three subdivisions are seen: a period of cell growth after the cell has divided called G_1, the replication of the chromosomes called **S (synthesis)** and lastly another period of growth called G_2. Figure 2**.**

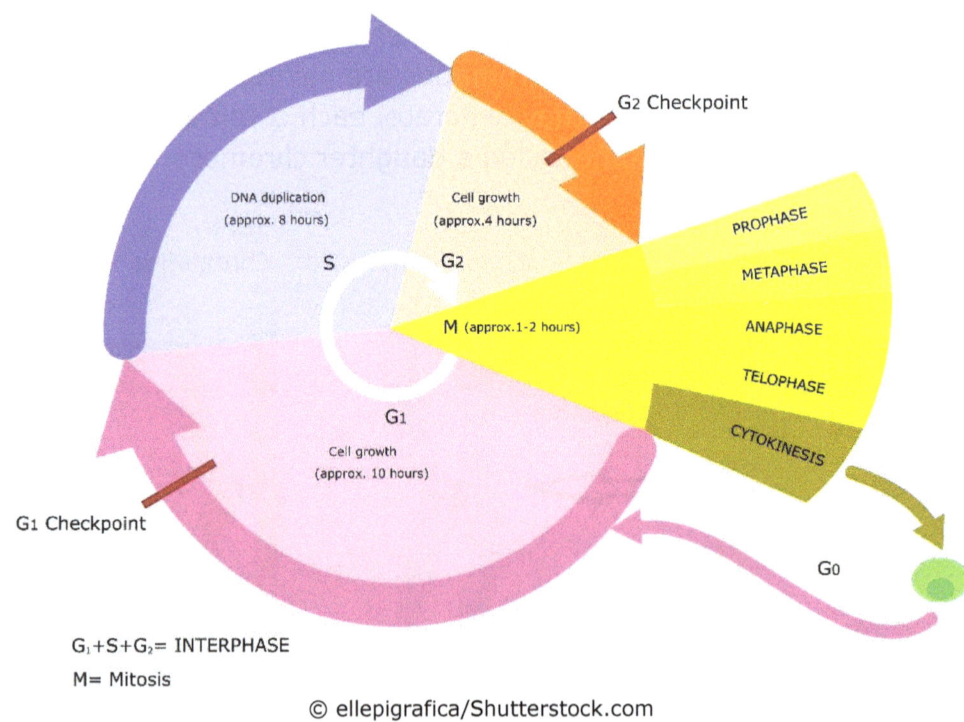

Figure 2 The Cell Cycle

Interphase (G₁, S, and G₂)

In interphase, the chromosomes replicate to form identical chromosomes and the nucleus is surrounded by the nuclear membrane. A pair of centrioles in the cytosol make a copy of themselves and microtubules start to extend outward from them.

Mitotic Cell Division (Mitosis)

<u>STAGES OF MITOSIS</u>

The stages of mitosis are: **prophase, prometaphase, metaphase, anaphase and telophase.** See Figure 3.

<u>Prophase</u>

After replication, the identical chromosomes pair up, with each member of the pair called sister chromatids joined at the centromere. The centrioles in the cytoplasm start moving to opposite poles and the spindle microtubules form.

<u>Prometaphase</u>

The nuclear envelope fragments and the spindle joins the centrioles now at the poles.

<u>Metaphase</u>

The chromosomes (made up of two sister chromatids) are now lined up at the center of the cell connected to the spindle.

<u>Anaphase</u>

The sister chromatids separate from each other and reach opposite ends of the cell. Each chromatid is now referred to as an individual chromosome.

<u>Telophase</u>

A nuclear envelope forms around the collection of chromosomes at each pole of the cell. Two nuclei are seen and <u>cytokinesis</u> begins.

Cytokinesis

In animal cells, a **cleavage furrow** forms dividing the parent cell into two daughter cells.

In plant cells, a **cell plate** forms to separate the parent cell into two daughter cells.

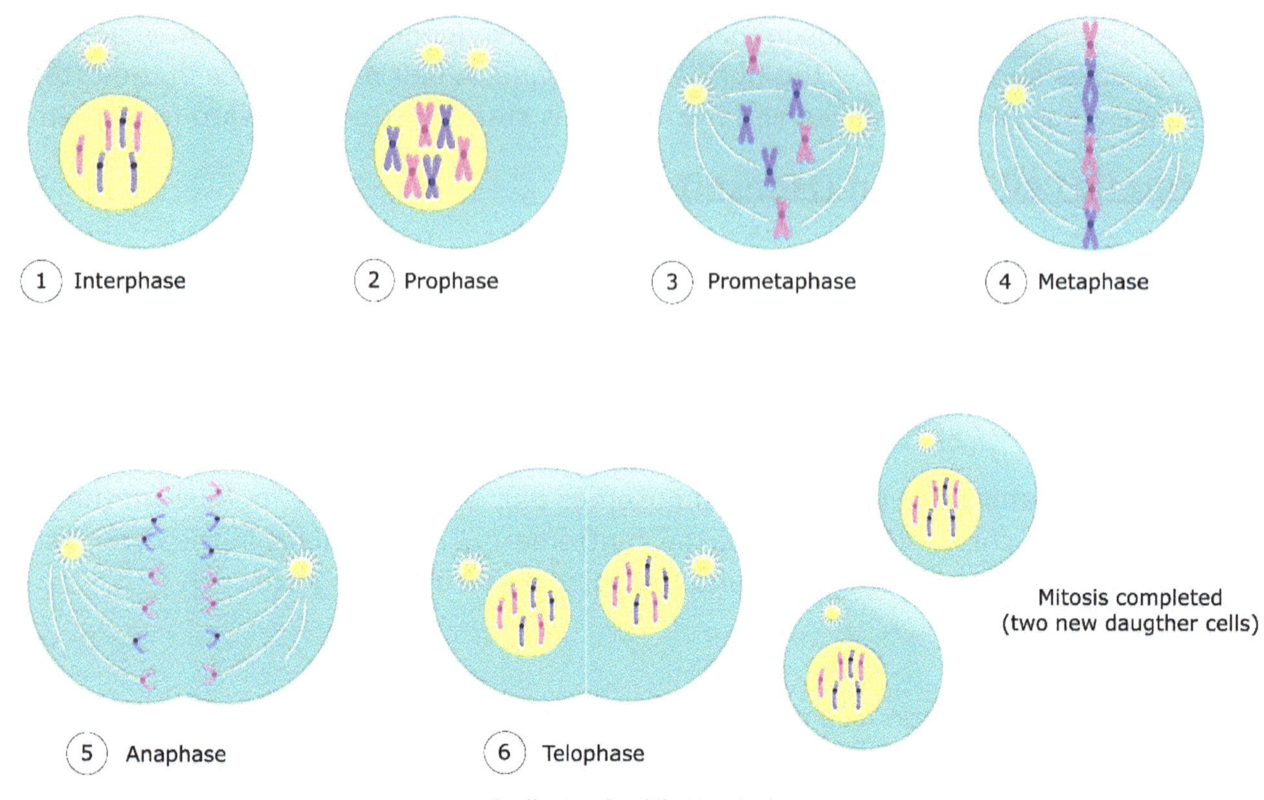

© ellepigrafica/Shutterstock.com

Figure 3 Stages of Animal Mitosis

<u>Procedure 1 Prepared Slide of Onion Root Tip</u>

The rapidly dividing cells of the tip of an onion root are excellent for studying the different stages of mitosis. Figure 5. You will be focusing on the region of cell division (the mitotic zone seen on the next page), where cells are actively dividing. Figure 4.

© Paper Teo/Shutterstock.com

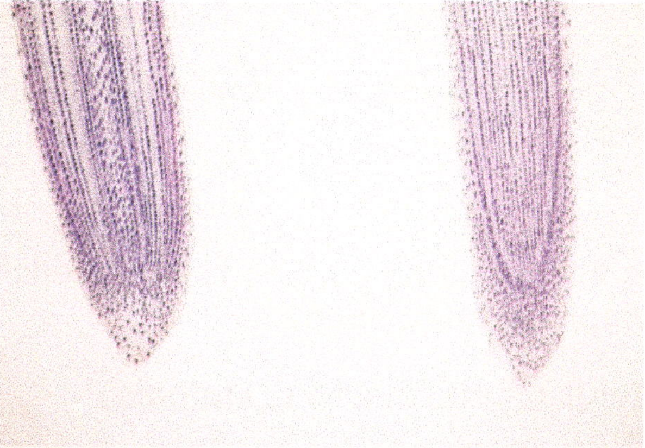

© Rattiya Thongdumhyu/Shutterstock.com

Figure 4 Onion Root Tips

1. Examine a prepared slide of an onion root tip.
2. Locate the cells near the tip of the root undergoing mitosis.
3. Choose a section of about 100 cells and count the number in interphase, prophase, metaphase, anaphase and telophase.
4. Record the results in the table on the next page.

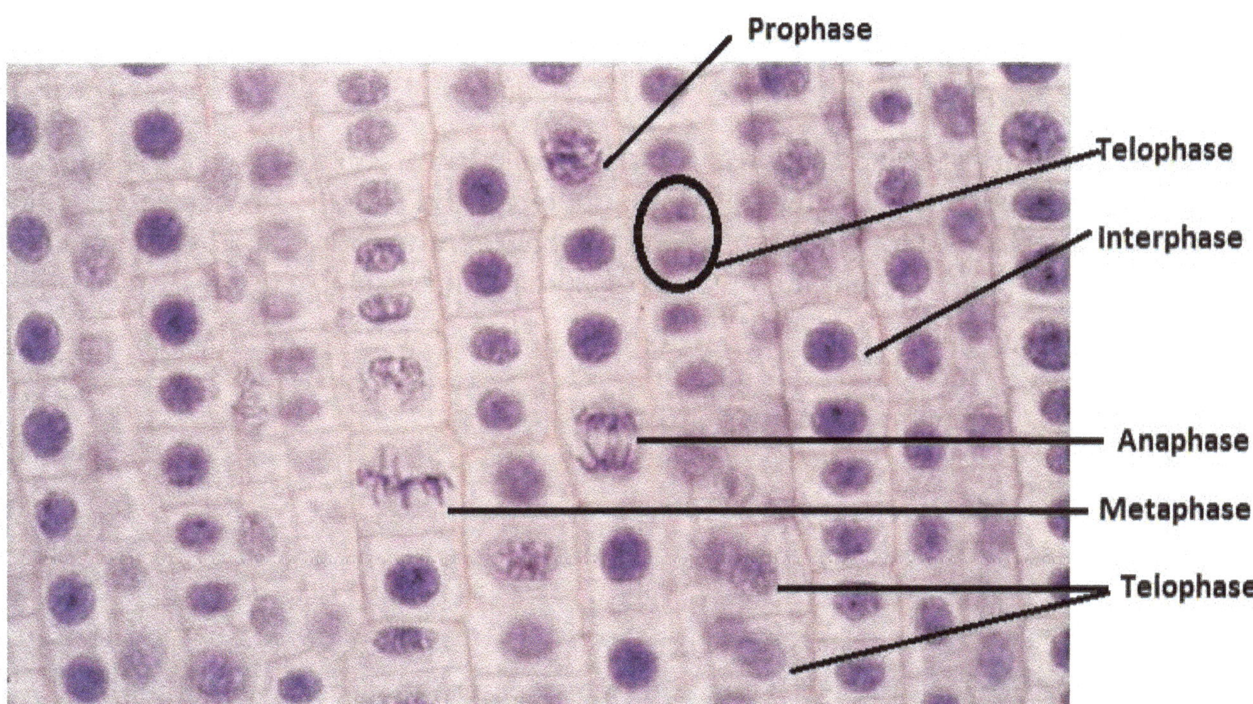

© Rattiya Thongdumhyu/Shutterstock.com/ Modified by Werner Williams
Figure 5 Onion Cells Showing Mitosis

125

Onion Root Tip Mitotic Stages

	Interphase	Prophase	Metaphase	Anaphase	Telophase
# of cells					

Source: Werner Williams

Be able to identify onion root tip cells in interphase, prophase, metaphase, anaphase and telophase.

Meiotic Cell Division (Meiosis)

Meiosis is different from mitosis in that meiosis has two divisions. Four daughter cells are produced with half the number of chromosomes of the diploid (2n) parent cell. So the daughter cells contain only one member of each chromosome pair and become the haploid (n) gametes (eggs and sperm).

Meiosis also reshuffles the genes which results in significant genetic variety in the daughter cells.

STAGES OF MEIOSIS (Figure 6 and 7)

Meiosis I

Prophase/prometaphase I

After the process of interphase, two sister chromatids form a pair called a **dyad** and then two dyads associate together in a group called a **tetrad**. The tetrad can also be referred to as a **pair** of **homologous chromosomes**. Crossing over may occur at this stage. The nuclear membrane fragments, and a spindle forms.

Metaphase I

The tetrads line up at the center of the cell.

Anaphase I

The dyads in each tetrad separate and start moving to opposite ends of the cell.

Telophase

The dyads arrive at the poles and a nuclear membrane forms around them.

Cytokinesis then occurs, dividing the parent cell into daughter cells.

Meiosis I

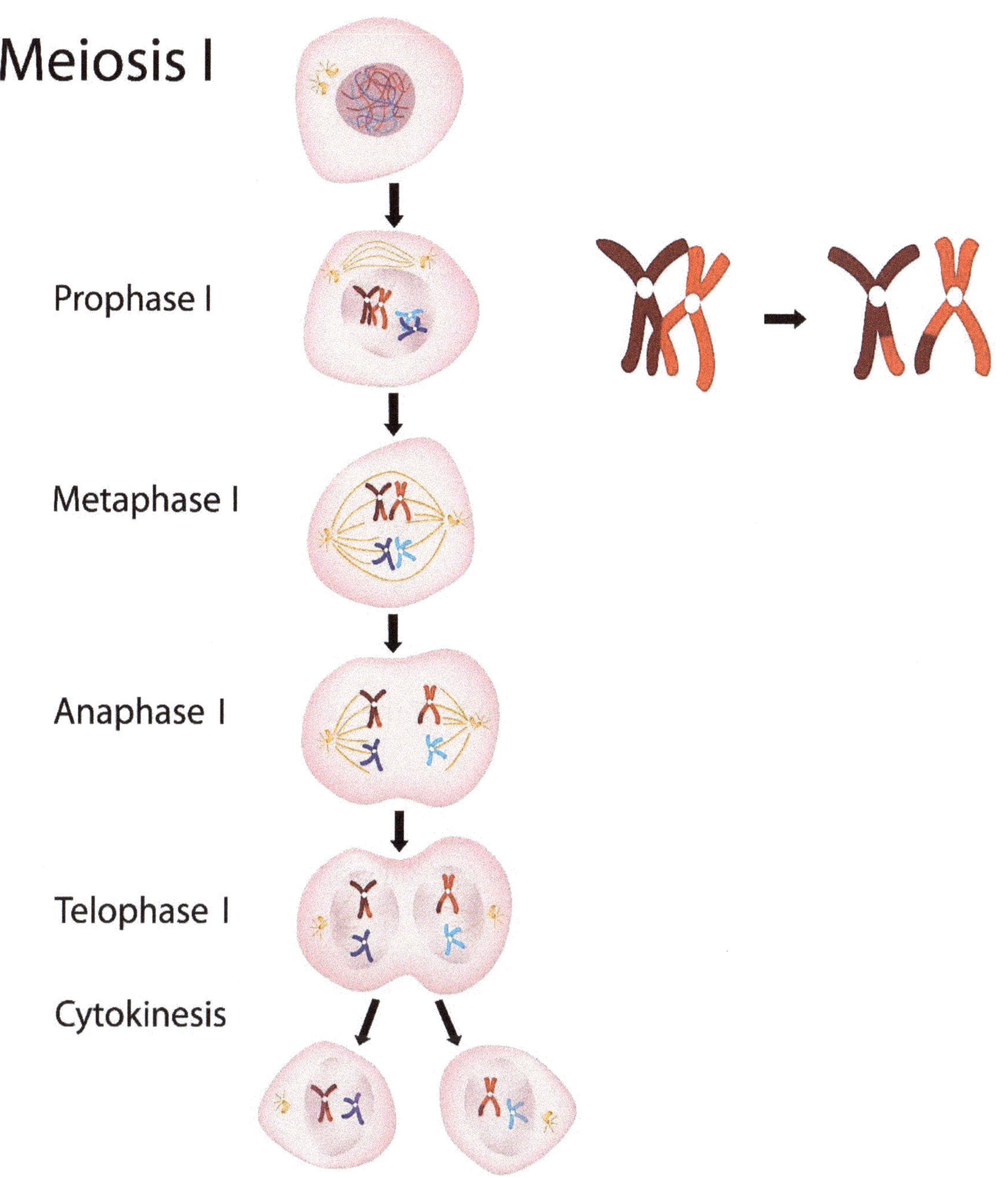

© Alila Medical Media/Shutterstock.com

Figure 6 Meiosis I

Meiosis II

Both cells that are formed in meiosis I divide again in meiosis II. The chromosomes do not replicate again because replication has already occurred (in meiosis I). Meiosis II is very similar to mitosis.

Prophase II

The centrioles start moving to opposite poles and the spindle microtubules form.

Prometaphase II

The nuclear envelope fragments. The spindle joins the centrioles now at the poles.

Metaphase II

The dyads are now lined up at the center of the cell connected to the spindle.

Anaphase II

The sister chromatids start to separate from each other (now called daughter

chromosomes), until they reach opposite ends of the cell.

Telophase II

The nuclear envelope forms around the collection of chromosomes at each pole of the cell. Two nuclei are seen in the cell and cytokinesis forms the daughter cells.

Meiosis II

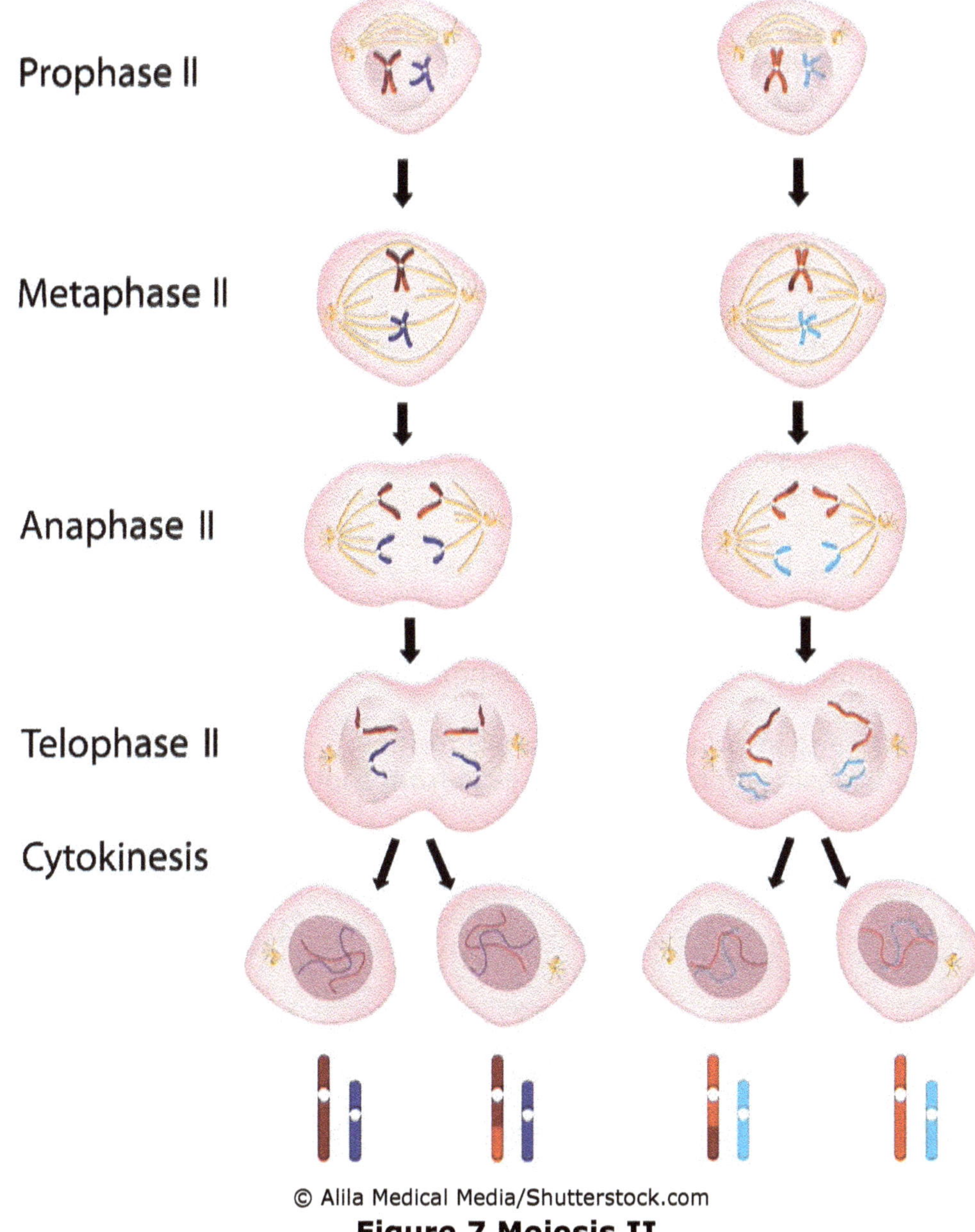

© Alila Medical Media/Shutterstock.com

Figure 7 Meiosis II

<u>Procedure 2 Mitosis and Meiosis Models</u>

Study the models of mitosis and meiosis.

129

<u>QUESTIONS</u>

Write the term that completes each phrase.

1. The haploid # of chromosomes in humans. _______________
2. The diploid # of chromosomes in humans. _______________
3. The mitotic stage in which the nuclear membrane disappears._______________
4. What would be the number of chromosomes in the parent cell of humans when considering meiosis? _______________________________________
5. What would be the number of chromatids in human daughter cells formed by the first division in meiosis? _______________________________________
6. What would be the number of chromosomes in human daughter cells formed by the first division in meiosis? _______________________________________
7. What would be the number of chromosomes in human daughter cells formed by the second division in meiosis? _______________________________________
8. What would be the number of human haploid cells formed by meiotic division of the parent cell? _______________________________________
9. The mitotic stage in which the daughter cells start to form._______________
10. The mitotic stage in which the chromosomes start separating and moving to opposite poles._______________________
11. The mitotic stage in which the chromosomes line up at the center of the cell._______________
12. The mitotic stage in which the spindle is formed._______________
13. How is cytokinesis different in plant cells from that in animal cells?

____

____

____

14. What is the division of the cytoplasm called?
____

15. How does anaphase I differ from anaphase in mitosis? _______________

____

____

____

<u>Lab 12: Human Genetics</u>

Humans have 46 chromosomes in their body cells. 44 of them are called **autosomes** and the other 2 are called the **sex chromosomes**. Females have 2X sex chromosomes and males have an X and a Y sex chromosome.

We have two genes for most of our biological characteristics, one gene from Dad and one from Mom. Letters are used to represent our genes, also called alleles, with a capital letter representing the dominant characteristics and a lower-case letter representing the recessive characteristic. If an individual's genotype is homozygous dominant (FF) or heterozygous (Ff), their phenotype will show the dominant characteristic. If an individual is recessive (ff), their phenotype will show the recessive characteristic.

Let's look at some common autosomal human traits. With your lab partner, determine each other's phenotypes from the characteristics listed in Table 1. Record your phenotype and probable genotype in the table. You may visualize these phenotypes on the next page.

Table 1 Autosomal Human Characteristics

Characteristic: d=dominant; r=recessive	Possible Genotypes	Your Phenotype	Your Genotype
Skin pigmentation: freckles (d); no freckles(r)	FF or Ff ff		
Hairline: widow's peak (d); straight hairline (r)	WW or Ww ww		
Bent little finger: little finger bends toward ring finger (d); straight little finger (r)	BB or Bb bb		
Thumb hyperextension: regular thumb (d); hitchhiker's thumb (r)	TT or Tt tt		
Earlobes: unattached (d); attached(r)	UU or Uu uu		

Source: Werner Williams

© Tracy Whiteside/Shutterstock.com

Freckles

© Malyugin/Shutterstock.com

Regular thumb

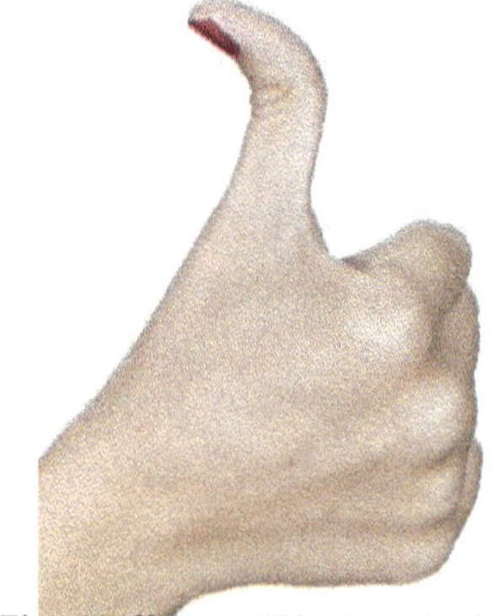

© Elsa Hoffmann/Shutterstock.com

Hitchhiker's thumb

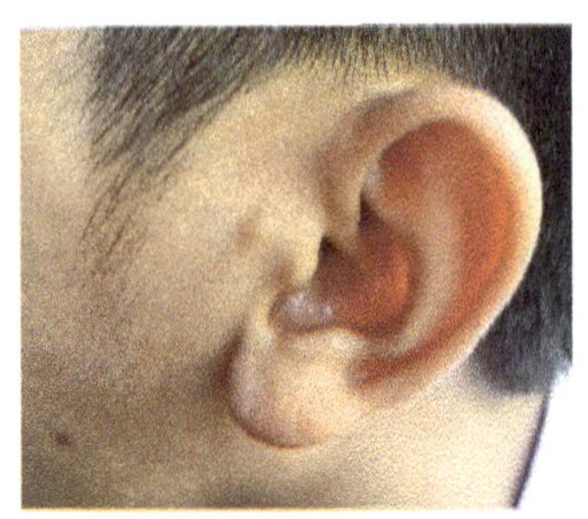

© Thiti Sukapan/Shutterstock.com

Unattached earlobes

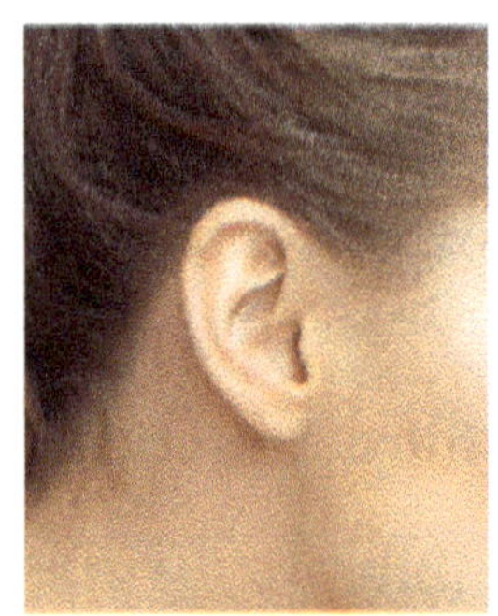

© Syda Productions Shutterstock.com

Attached earlobes

© Everett Collection/Shutterstock.com

Widow's peak

© chombosan/Shutterstock.com

Straight hairline

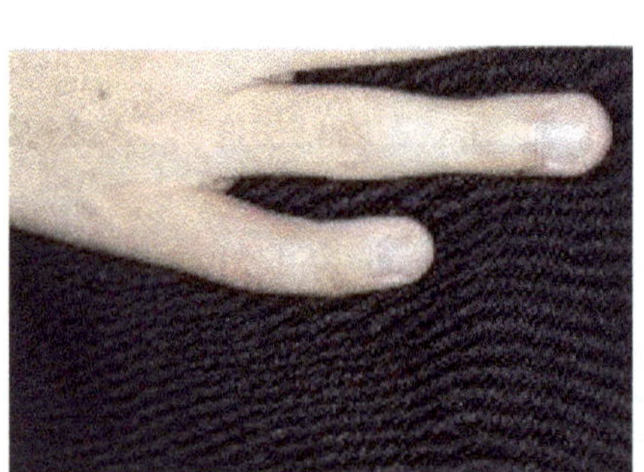

© Dr P. Marazzi / Science Source

Bent little finger

© Monster Ztudio/Shutterstock.com

Straight little finger

1. Mike's parents have freckles but Mike does not. What is the expected percentage among the children in this family? % with freckles_________; % without freckles______________

2. Susan is adopted. She has hitchhiker's thumb. Could both of her parents have hitchhiker's thumb? _________________ Could both of her parents not have hitchhiker's thumb?_________Explain.

3. Janelle and everyone in her immediate family have straight hairlines. Her maternal grandmother has widow's peak. What is the genotype of her maternal grandmother? ______
Her maternal grandfather is no longer alive. What could have been his genotype?_________

Tasting PTC

Some people have the ability to taste PTC (phenylthiocarbamide) which is bitter, while others cannot taste it. Tasting (T) is dominant to non-tasting (t).

<u>Procedure 1</u>

1. Taste a piece of paper that contains PTC. Can you taste it?_______________
2. What is your probable genotype (s)? _________________________________

Genetics Problem

Jordan's parents can taste PTC but Jordan cannot taste it. Construct a punnet square to show the probable genotypes and phenotypes of the possible children in that family.

ABO Blood Types

There are three genes that control the ABO blood types: A, B and O. A and B are dominant over O and A and B are co-dominant to each other. Each person has 2 of the 3 possible genes.

Table 2 Blood Types

Genotypes	Phenotypes (Blood Type)
AA or AO	A
BB or BO	B
AB	AB
OO	O

Source: Werner Williams

<u>Genetics Problems</u>

1. Could a man with blood type B and a woman with type AB produce a child with type O blood? _____________________________

2. If the mother is blood type O and the father is A, what could be the blood type(s) of their children ?____

3. A boy wonders if he is adopted. If the father is blood type AB and the mother is blood type O, what blood type would indicate that the child might have been adopted? Blood type________________

Colorblindness

Colorblindness is an example of a sex-linked recessive characteristic. Sex-linked means the gene is found on one of the sex chromosomes (with color-blindness, it is found on the X chromosome). If the person has a recessive genotype, the person would be colorblind. Consider the following genotypes and phenotypes of various individuals:

Females

X^BX^B (normal vision)

X^BX^b (normal vision)

X^bX^b (color-blind)

Males

X^BY (normal vision)

X^bY (color-blind)

Genetics Problems

1. Rianna and her father are both color-blind but her mother has normal vision. What are the genotypes of these members of the family?
Rianna____________; Dad_________;Mom ________________

2. Rianna's sister Caroline has normal vision. What is her genotype?
____

3. The only colorblind member of Bob's family is his brother Jacob. What is Jacob's genotype?________________________What is Bob's father's genotype?
____

4. What is Bob's mother's genotype?_______________________________If Bob's sister
   Mary has a colorblind son, what is Mary's genotype?______________________________

Pedigrees

Pedigrees show the inheritance of a genetic disorder within a family and
allows you to determine whether any particular person in that family has the
gene for that disorder. Pedigrees show the person's phenotype. We can use
this information to determine what are the chances that a particular couple
would produce a normal or affected child.

(A carrier is a person who does not show the affected characteristic but may
transmit the characteristic to their children).

Figure 1 shows you the symbols used in a typical pedigree chart.

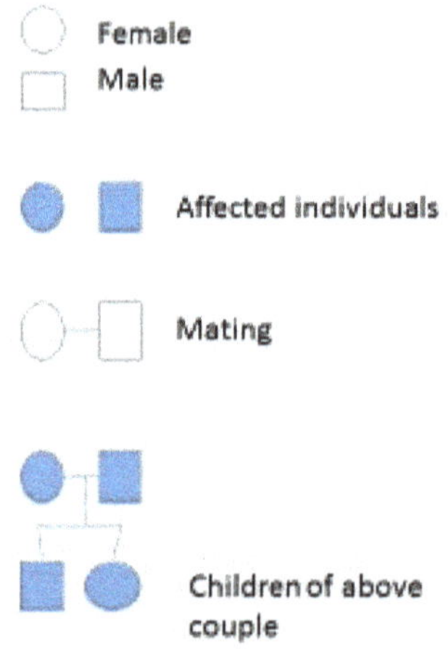

Source: Werner Williams
Figure 1 Pedigree Symbols

For each of the three pedigrees below, determine if the disorder or
characteristic is autosomal dominant, autosomal recessive or X-linked
recessive **and write the person's genotype beside their circle or square**.
Use Table 3 to help you.

Table 3 Pedigree Solution

The Key That Solves The Pedigree	Possible Genotypes
Autosomal Dominant	AA or Aa= affected; aa= normal
Autosomal Recessive	AA or Aa= normal; aa= affected
X-linked Recessive	$X^A X^A$ and $X^A X^a$ = normal female; $X^a X^a$ = affected female; $X^A Y$ = normal male; $X^a Y$ = affected male

Source: Werner Williams

Problems

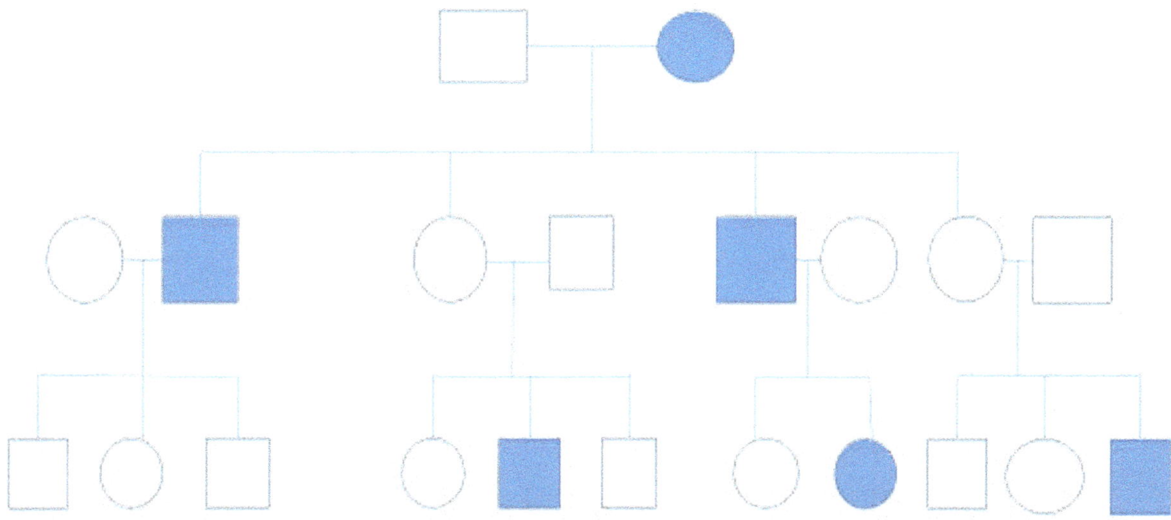

Source: Werner Williams

Key to this pedigree: ______________________________________

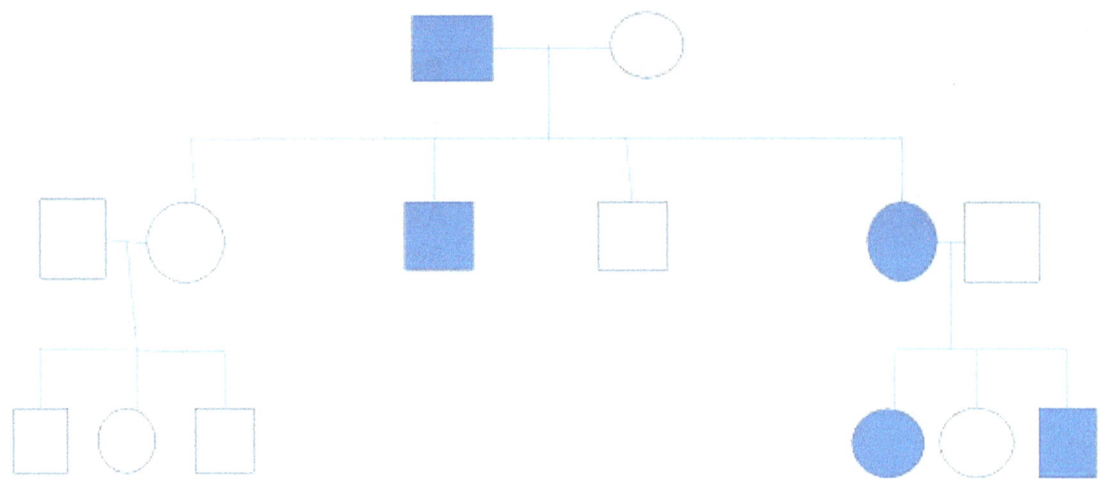

Source: Werner Williams

Key to this pedigree:____

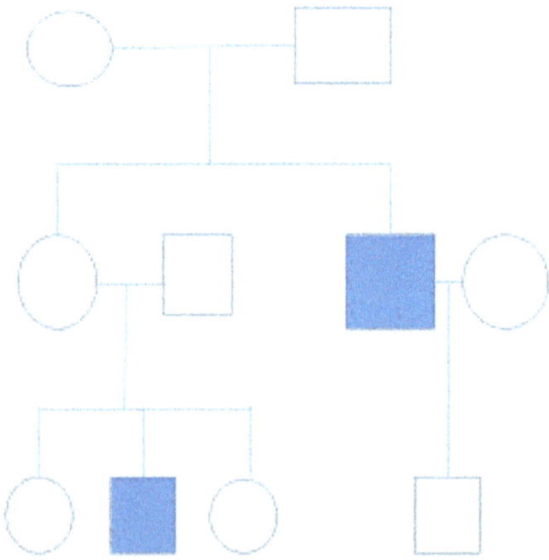

Source: Werner Williams

Key to this pedigree:____